Introduction to Tribology

Introduction to Tribology

Elden Davidson

STATES
ACADEMIC PRESS
www.statesacademicpress.com

States Academic Press,
109 South 5th Street,
Brooklyn, NY 11249, USA

ISBN: 978-1-63989-314-0

Cataloging-in-Publication Data

Introduction to tribology / Elden Davidson.
p. cm.
Includes bibliographical references and index.
ISBN 978-1-63989-314-0
1. Tribology. 2. Friction. 3. Surfaces (Technology). I. Davidson, Elden.
TJ1075 .I58 2022
621.89--dc23

For information on all States Academic Press publications
visit our website at www.statesacademicpress.com

STATES
ACADEMIC PRESS

TABLE OF CONTENTS

PREFACE

This book aims to help a broader range of students by exploring a wide variety of significant topics related to this discipline. It will help students in achieving a higher level of understanding of the subject and excel in their respective fields. This book would not have been possible without the unwavered support of my senior professors who took out the time to provide me feedback and help me with the process. I would also like to thank my family for their patience and support.

The science and engineering of interacting surfaces in relative motion is referred to as tribology. The field focuses on the study and application of concepts of friction, wear and lubrication. The force that resists the rolling and sliding of one solid object over another is known as friction. Wear refers to the damage and gradual removal and deformation of materials from solid surfaces. Its causes can be chemical and mechanical. Lubrication is the process and the technique of using lubricants to reduce wear and tear, and friction between two surfaces. Primary branches of this discipline include classical tribology, biotribology, green tribology, geotribology, nanotribology and tribotronics. Other important branches are computational tribology, open system tribology and space tribology. This book is a compilation of chapters that discuss the most vital concepts in this field. Most of the topics introduced herein cover new techniques and applications of tribology. Those in search of information to further their knowledge will be greatly assisted by this book.

A brief overview of the book contents is provided below:

Chapter - Introduction

Engineering tribology is an interdisciplinary subject that deals with the study of interaction of surfaces in relative motion. It mainly covers three major concepts of friction, wear and lubrication. This chapter has been carefully written to provide an easy understanding of engineering tribology.

Chapter - Friction and Wear

Friction is the resistance against the motion of solid surfaces, fluid layers, and material elements. A few of its types are dry friction, fluid friction, skin friction, internal friction, etc. Wear is the damaging or deformation of materials at solid surfaces. This chapter discusses the subject of friction and wear associated with engineering tribology in detail.

Chapter - Lubrication and Lubricants

Lubrication can be defined as the process in which lubricants are used for reducing friction and wear between two surfaces. A few types of lubricants include mineral and synthetic oils, biolubricants, solid lubricants, etc. This chapter closely examines lubrication and lubricants to provide an extensive understanding of the subject.

Chapter - Bearings and Its Types

A bearing is an element of machine that resists its motion and helps in reducing the friction between moving parts. Hydrodynamic thrust bearings use hydrodynamic lubrication for positioning and movement of axis load from shift to stationary part. All the aspects related to bearings and hydrodynamic thrust bearings have been carefully analyzed in this chapter.

Chapter - Tribological Aspects of Gears

There are many aspects that are dealt within the tribology of gears such as spur gears, gear configurations, contact mechanics, monitoring system of gears, gear failure and its management, etc. The topics elaborated in this chapter will help in gaining a better perspective of the tribological aspects of gears.

Elden Davidson

Introduction 1

- **Tribology**
- **Applications of Tribology**

Engineering tribology is an interdisciplinary subject that deals with the study of interaction of surfaces in relative motion. It mainly covers three major concepts of friction, wear and lubrication. This chapter has been carefully written to provide an easy understanding of engineering tribology.

Tribology

Tribology is the study of the interaction of sliding surfaces. It includes three subjects: friction, wear, and lubrication. There is a difficulty in that friction is generally characterized as a branch of physics or mechanical engineering wear is part of the material science of metallurgy, while lubrication is a branch of chemistry. Tribology is thus a complex interdisciplinary subject.

The phenomena considered in tribology are among the most fundamental and most common of those encountered by humans in interaction with their largely solid environment. Many manifestations of tribology are beneficial and, indeed, make modern life possible. Many other effects of tribology, however, constitute serious nuisances, and careful design is necessary to overcome the inconvenience arising from excessive friction or wear. On an overall basis, friction uses up, or wastes, a substantial amount of the energy generated by mankind, while a large amount of productive capacity is devoted to replacing objects made useless by wear.

Friction is the resistance to sliding of a solid when the resistance is produced by a contacting body. It is therefore a vital factor in the operation of most mechanisms. High friction is needed for the satisfactory functioning of nuts and bolts, paper clips, and tongs, as well as in the familiar processes of walking, gripping objects manually, and building piles of sand or of apples. Low friction, however, is desired in objects that are designed to move continuously, like engines, skis, and the internal mechanism of watches. Constant friction is required in brakes and clutches, as otherwise unpleasant jerky movement would be produced.

Friction has been studied as a branch of mechanics for many hundreds of years and its laws, as well as satisfactory methods of estimating the magnitude of friction, have been known for nearly two centuries. The mechanism of friction, namely, the exact process by which energy is lost as two surfaces slip past each other, is understood only in an incomplete way.

Wear is the removal of material from a solid surface as a result of the mechanical action exerted by another solid. It is such a universal phenomenon that rarely do two solid bodies slide over each other or even touch each other without measurable material transfer or material loss. Thus, coins become worn as a result of continued contact with human fingers; pencils become worn after sliding over paper; and rails become worn as a result of the continued rolling of train wheels over them. Only living things (e.g., bone joints) are in general immune to the permanent damage caused by wear because only they have the property of healing through regrowth. And even a few living things do not heal themselves (e.g., teeth in humans).

The systematic study of wear has been severely hampered by two factors: first, the existence of a number of separate wear processes, which has led to much confusion, especially in terminology; second, the difficulties caused by the small amounts of material involved in wear processes. These difficulties were greatly alleviated when radioactive isotopes of the common engineering metals (iron, copper, chromium, etc.) became available in the 1940s; tracer techniques using these radioisotopes permit measurements of wear, even in small amounts, while it is occurring. This has made it possible to identify types of wear and discover the laws of wear.

The use of lubricants, namely, substances introduced into the interface between sliding surfaces to diminish friction, is an ancient practice, and Egyptian pictures dating back 4,000 years show the application of lubricants to reduce the friction involved in dragging heavy monuments. In modern lubrication practice, the main concern is to reduce the wear that accompanies sliding and, at the same time, to design lubrication systems that will operate for long periods without inspection or maintenance.

A large number of different lubricants are in use at any one time, (a single major oil company may market many hundreds of different varieties), and no aspect of tribology receives as much attention as the development and testing of improved lubricants.

Applications of Tribology

Biomedical (Biotribology)

The application of tribology in biological systems is a rapidly growing field and extends well beyond the conventional boundaries. Biomedical tribological systems involve an extensive range of synthetic materials and natural tissues, including cartilage, blood vessels, heart, tendons, ligaments, and skin.

These materials operate in complex interactive biological environments. Biotribologists incorporate concepts of friction, wear, and lubrication of these biological surfaces in various applications, such as the design of joints and prosthetic devices, the wear of screws and plates in bone fracture repair, wear of denture and restorative materials, wear of replacement heart valves, and even the tribology of contact lenses.

Nanotribology

The commercialization of microelectromechanical systems (MEMS)/nanoelectromechanical systems (NEMS), such as disk drives and other magnetic storage systems in the early 1990s, along with the development of new materials with nanoscale thicknesses, have presented new tribological challenges.

The emergence of sophisticated scanning probe technologies and computational techniques has given rise to the field of nanotribology for investigations of processes at the atomic,molecular, andmicroscopic scale. Nanotribological studies are helping to develop fundamental understanding of surface interfaces in micro/nanostructures used in a variety of modern applications.

Some of these applications include chemical and biodetectors, advanced drug delivery systems, information recording layers, molecular sieves, systems on a chip, nanoparticle reinforced materials, and a new generation of lasers.

Wind Turbines

Wind turbines continue to hold promise as a viable alternative energy source. While they have made gains in reliability in the past decade, they are subject to tribological problems that are difficult and costly to repair and can drastically reducing their expected lifecycles. Two key areas of concern are reliability of gearboxes and turbine lubrication.

To address these problems, there have been extensive studies of lubricants, which are used in wind-turbine gearboxes, hydraulic systems such as the blade pitch control, drive train brake subsystems, and bearings, to improve turbine reliability in extreme environments.

Green Tribology

The concept of "green tribology" was also introduced by Jost, who defined it as, "The science and technology of the tribological aspects of ecological balance and of environmental and biological impacts."

There are a number of problems that can be addressed by green tribology. The specific field of green or environment-friendly tribology emphasizes the aspects of interacting surfaces in relative motion, which are of importance for energy or environmental sustainability or which have impact upon today's environment.

Nosonovsky and Bhushan suggested the 12 principles of green tribology as the minimization of (1) friction and (2) wear, (3) the reduction or complete elimination of lubrication, including self-lubrication, (4) natural and (5) biodegradable lubrication, (6) using sustainable chemistry and engineering principles, (7) biomimetic approaches, (8) surface texturing, (9) environmental implications of coatings, (10) real-time monitoring, (11) design for degradation, and (12) sustainable energy applications.

They further emphasized three areas of green tribology: Biomimetics for tribological applications, environment-friendly lubrication, and the tribology of renewable-energy application.

Friction and Wear 2

- Friction
- Laws of Friction
- Dry Friction and its Causes
- Laws of Dry Friction
- Wear
- Theories of Wear
- Wear Prevention
- Wear Testing and Wear Measurement

Friction is the resistance against the motion of solid surfaces, fluid layers, and material elements. A few of its types are dry friction, fluid friction, skin friction, internal friction, etc. Wear is the damaging or deformation of materials at solid surfaces. This chapter discusses the subject of friction and wear associated with engineering tribology in detail.

Friction

Friction is a force acting opposite to the direction of relative motion. Friction rises on the interface between bodies, but may also develop within the body. Examples of the latter include air and hydrodynamic friction, where the friction rises between the fluid layers. Friction is not a fundamental force, but rather a manifestation of electro-magnetic and gravity forces. There are many sources of friction: deformation of the surfaces, adhesion, capillary forces, van der Waals forces, chemical bonding, etc. Friction is a non-conservative force and the energy spent to overcome friction is lost. Friction losses reduce the energy efficiency of many mechanical devices. A recent report on the energy consumption due to frictional losses in passenger cars estimates that only 21.5% of the potential fuel energy is used to move the car, whereas direct frictional losses account for 33% (28% if breaking is excluded). Within these losses, 35% is used to overcome tire-road friction, 35% to overcome engine friction, 15% – transmission and 15% brake contact friction. So one of the major challenges of the modern tribology is to find new technologies to reduce friction and increase energy efficiency.

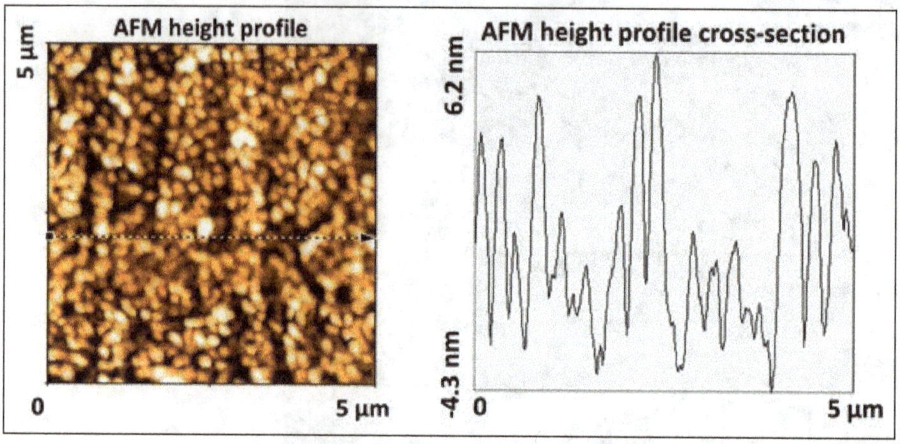

AFM height profile and cross-section of clean glass plate.

In general, friction can be dry or lubricated. For a vast majority of the surfaces encountered in nature and used in industry, the source of dry friction is the imperfections of the surfaces. It turns out that even mirror shining surfaces are actually not flat, at least at micro- and nano-scales, and contain hills and valleys. These imperfections are called surface roughness. A typical Atomic Force Microscopy measured surface profile of a smooth clean glass plate is shown below.

Egg carton type surfaces.

When this rough surface is dragged against another rough surface, these valleys and hills will interlock with each other, deform and consequently oppose the motion. The classical example of the interlocking are the stalled egg cartons type surfaces, which are hard to slide against each other (due to high friction) as shown in the figure below. The force needed to overcome this interlocking is friction.

At the same time, producing atomically flat surfaces does not solve the problem of friction reduction. Actually the decrease of roughness to atomically flat level for engineering materials, such as steels, for example, leads to a much higher dry friction. And this is due to an increase of the adhesion forces (van der Waals interactions, capillary forces, etc).

In order to reduce friction the lubrication theory is used. This approach introduces a lubricant film between rough surfaces and the direct contact between rough surfaces is either eliminated or minimized. The friction in this case is determined mostly by the shear of lubricant layers, rather than shearing of solids, as shown in the figure below.

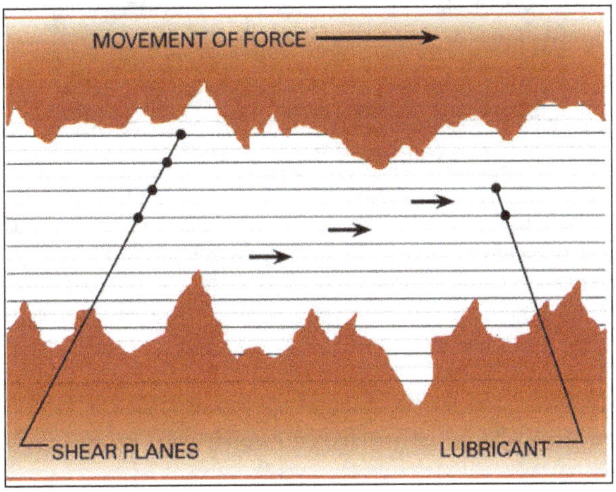

Lubricant Film.

In such case, the interlocking and adhesion effects are reduced. For most engineering materials, dry friction is larger than lubricated since the shear strength of the solids is typically larger. However, the theory has reached its fundamental limit (0.01-0.04 friction coefficient) and cannot be used to reduce the friction further. Therefore, the researchers look for other ways to reduce friction and recently several cases of superlubricity (friction coefficient well below 0.01) were reported.

Classification of Friction

Different types of motion of the object gives rise to different types of friction. Generally, there are four types of friction. They are static friction, sliding friction, rolling friction, and fluid friction.

Static Friction

Static friction exists between a stationary object and the surface on which it is resting.

It prevents an object from moving against the surface. Example: Static friction prevents an object like a book from falling of the desk, even if the desk is slightly tilted. It helps us pick up an object without it slipping through our fingers.

A stationary book on surface.

When we want to move an object first we must overcome the static friction acting between the object and the surface on which the object is resting.

Sliding Friction

Sliding friction occurs between objects as they slide against each other. When sliding friction is acting there must be another force existing to keep the body moving.

Example: When a man is pushing an object on a rough surface the force acting is called "sliding friction".

Rolling Friction

Rolling friction is the resistive force that slows down the motion of a rolling ball or wheel. It is also called rolling resistance.

When a force or torque is applied to a stationary wheel, there is a small static rolling friction force holding back the rolling motion. However, resistance from static sliding friction is what really causes the wheel to start rolling. Rolling friction hinders the motion of an object rolling on a surface that means it slows down the motion of an object rolling on a surface. Examples: It slows down a ball rolling on a surface and it slows down the motion of tire rolling on the surface.

Like sliding friction here also another force is required to keep the object in motion, in case of pedaling bicycle the bicyclist provides the force which is required for the bicycle to be in motion.

A ball rolling on a surface.

Fluid Friction

Here on Earth we tend to take air resistance ("drag") for granted. We just assume that when we throw a ball, launch an aircraft, deorbit a spacecraft, or fire a bullet from a gun, that the act of it traveling through our atmosphere will naturally slow it down. But what is the reason for this? Just how is air able to slow an object down, whether it is in free-fall or in flight? Air friction is experienced by the objects moving through the open air. Air friction acts between the object and the air through which it is moving. It is also called drag. This force depends upon the object's shape, material, speed with which it is moving and the viscosity of the fluid. Viscosity is the measure of the resistance of the air to flow and it differs from one density another.

Example: It slowdowns the motion of airplane flying in the air, here the engine of the airplane helps the plane to overcome the fluid friction and move forward.

A plane in the air.

Laws of Friction

- Friction depends on the hardness or roughness of the contacting surfaces. Fluid friction depends on the viscosity (the thickness) of the fluid.
- Friction is directly proportion to the normal force pressing the contacting forces together.

- Friction does not depend on the area of the contacting surfaces.
- In the case of sliding, friction is reduced at very high relative speeds. However, in the case of fluid friction, friction increases with increase in relative speed of movement.

To stop an object in motion, a force must act on it in the opposite direction of motion. The force that opposes the motion of the object is called the frictional force.

Look at the diagram. At first the block is at rest, then the pushing force keeps the block moving. As the block slides over the surface, the frictional force acts on it in the opposite direction. A unit of friction is a Newton, as forces are measured using Newtons.

Friction generally depends on weight of the object and nature of the surface between the moving object and supporting surface.

Dry Friction and its Causes

Dry friction resists relative lateral motion of two solid surfaces in contact. The two regimes of dry friction are 'static friction' ("stiction") between non-moving surfaces, and kinetic friction (sometimes called sliding friction or dynamic friction) between moving surfaces.

Coulomb friction, named after Charles-Augustin de Coulomb, is an approximate model used to calculate the force of dry friction. It is governed by the model:

$$F_f \leq \mu F_n,$$

where,

- F_f is the force of friction exerted by each surface on the other. It is parallel to the surface, in a direction opposite to the net applied force.
- μ is the coefficient of friction, which is an empirical property of the contacting materials.
- F_n is the normal force exerted by each surface on the other, directed perpendicular (normal) to the surface.

The Coulomb friction F_f may take any value from zero up to μF_n, and the direction of the frictional force against a surface is opposite to the motion that surface would experience in the absence of friction. Thus, in the static case, the frictional force is exactly what it must be in order to prevent motion between the surfaces; it balances the net force tending to cause such motion. In this case, rather than providing an estimate of the actual frictional force, the Coulomb approximation provides a threshold value for this force, above which motion would commence. This maximum force is known as traction.

The force of friction is always exerted in a direction that opposes movement (for kinetic friction) or potential movement (for static friction) between the two surfaces. For example, a curling stone sliding along the ice experiences a kinetic force slowing it down. For an example of potential movement, the drive wheels of an accelerating car experience a frictional force pointing forward; if they did not, the wheels would spin, and the rubber would slide backwards along the pavement. It is not the direction of movement of the vehicle they oppose, it is the direction of (potential) sliding between tire and road.

Normal Force

Free-body diagram for a block on a ramp. Arrows are vectors indicating directions and magnitudes of forces. N is the normal force, mg is the force of gravity, and F_f is the force of friction.

The normal force is defined as the net force compressing two parallel surfaces together, and its direction is perpendicular to the surfaces. In the simple case of a mass resting on a horizontal surface, the only component of the normal force is the force due to gravity, where $N = mg$. In this case, the magnitude of the friction force is the product of the mass of the object, the acceleration due to gravity, and the coefficient of friction. However, the coefficient of friction is not a function of mass or volume; it depends only on the material. For instance, a large aluminum block has the same coefficient of friction as a small aluminum block. However, the magnitude of the friction force itself depends on the normal force, and hence on the mass of the block.

If an object is on a level surface and the force tending to cause it to slide is horizontal, the normal force N between the object and the surface is just its weight, which is equal to its mass multiplied by the acceleration due to earth's gravity, g. If the object is on a tilted surface such as an inclined plane, the normal force is less, because less of the force of gravity is perpendicular to the face of the plane. Therefore, the normal force, and ultimately the frictional force, is determined using vector analysis, usually via a free body diagram. Depending on the situation, the calculation of the normal force may include forces other than gravity.

Coefficient of Friction

The coefficient of friction (COF), often symbolized by the Greek letter μ, is a dimensionless scalar value which describes the ratio of the force of friction between two bodies and the force pressing them together. The coefficient of friction depends on the materials used; for example, ice on steel has a low coefficient of friction, while rubber on pavement has a high coefficient of friction. Coefficients of friction range from near zero to greater than one. It is an axiom of the nature of friction between metal surfaces that it is greater between two surfaces of similar metals than between two surfaces of different metals— hence, brass will have a higher coefficient of friction when moved against brass, but less if moved against steel or aluminum.

For surfaces at rest relative to each other $\mu = \mu_s$, where μ_s is the coefficient of static friction. This is usually larger than its kinetic counterpart. The coefficient of static friction exhibited by a pair of contacting surfaces depends upon the combined effects of material deformation characteristics and surface roughness, both of which have their origins in the chemical bonding between atoms in each of the bulk materials and between the material surfaces and any adsorbed material. The fractality of surfaces, a parameter describing the scaling behavior of surface asperities, is known to play an important role in determining the magnitude of the static friction.

For surfaces in relative motion $\mu = \mu_k$, where μ_k is the coefficient of kinetic friction. The Coulomb friction is equal to F_f, and the frictional force on each surface is exerted in the direction opposite to its motion relative to the other surface.

Arthur Morin introduced the term and demonstrated the utility of the coefficient of friction. The coefficient of friction is an empirical measurement – it has to be measured experimentally, and cannot be found through calculations. Rougher surfaces tend to have higher effective values. Both static and kinetic coefficients of friction depend on the pair of surfaces in contact; for a given pair of surfaces, the coefficient of static friction is usually larger than that of kinetic friction; in some sets the two coefficients are equal, such as teflon-on-teflon.

Most dry materials in combination have friction coefficient values between 0.3 and 0.6. Values outside this range are rarer, but teflon, for example, can have a coefficient as low as 0.04. A value of zero would mean no friction at all, an elusive property. Rubber in contact with other surfaces can yield friction coefficients from 1 to 2. Occasionally it is maintained that μ is always < 1, but this is not true. While in most relevant applications μ < 1, a value above 1 merely implies that the force required to slide an object along the surface is greater than the normal force of the surface on the object. For example, silicone rubber or acrylic rubber-coated surfaces have a coefficient of friction that can be substantially larger than 1.

While it is often stated that the COF is a "material property," it is better categorized as a "system property." Unlike true material properties (such as conductivity, dielectric constant, yield strength), the COF for any two materials depends on system variables like temperature, velocity, atmosphere and also what are now popularly described as aging and deaging times; as well as on geometric properties of the interface between the materials, namely surface structure. For example, a copper pin sliding against a thick copper plate can have a COF that varies from 0.6 at low speeds (metal sliding against metal) to below 0.2 at high speeds when the copper surface begins to melt due to frictional heating. The latter speed, of course, does not determine the COF uniquely; if the pin diameter is increased so that the frictional heating is removed rapidly, the temperature drops, the pin remains solid and the COF rises to that of a 'low speed' test.

Approximate Coefficients of Friction

Materials		Static Friction,		Kinetic/Sliding Friction,	
Dry and clean		Lubricated	Dry and clean	Lubricated	
Aluminium	Steel	0.61		0.47	
Aluminum	Aluminum	1.05-1.35	0.3	1.4-1.5	
Gold	Gold			2.5	
Platinum	Platinum	1.2	0.25	3.0	
Silver	Silver	1.4	0.55	1.5	
Alumina ceramic	Silicon Nitride ceramic				0.004 (wet)
BAM (Ceramic alloy $AlMgB_{14}$)	Titanium boride (TiB_2)	0.04–0.05	0.02		
Brass	Steel	0.35-0.51	0.19	0.44	
Cast iron	Copper	1.05		0.29	
Cast iron	Zinc	0.85		0.21	
Concrete	Rubber	1.0	0.30 (wet)	0.6-0.85	0.45-0.75 (wet)
Concrete	Wood	0.62			
Copper	Glass	0.68			
Copper	Steel	0.53		0.36	
Glass	Glass	0.9-1.0		0.4	
Human synovial fluid	Cartilage		0.01		0.003
Ice	Ice	0.02-0.09			
Polyethene	Steel	0.2	0.2		
PTFE (Teflon)	PTFE (Teflon)	0.04	0.04		0.04
Steel	Ice	0.03			
Steel	PTFE (Teflon)	0.04-0.2	0.04		0.04
Steel	Steel	0.74-0.80	0.16	0.42-0.62	
Wood	Metal	0.2–0.6	0.2 (wet)		
Wood	Wood	0.25–0.5	0.2 (wet)		

Under certain conditions some materials have very low friction coefficients. An example is (highly ordered pyrolytic) graphite which can have a friction coefficient below 0.01. This ultralow-friction regime is called superlubricity.

Static Friction

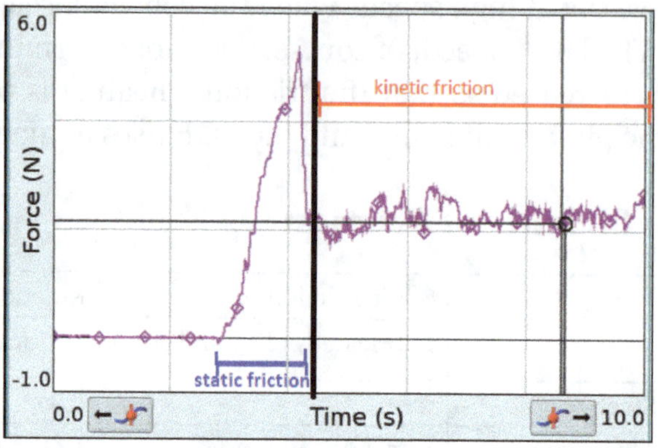

When the mass is not moving, the object experiences static friction. The friction increases as the applied force increases until the block moves. After the block moves, it experiences kinetic friction, which is less than the maximum static friction.

Static friction is friction between two or more solid objects that are not moving relative to each other. For example, static friction can prevent an object from sliding down a sloped surface. The coefficient of static friction, typically denoted as μ_s, is usually higher than the coefficient of kinetic friction. Static friction is considered to arise as the result of surface roughness features across multiple length-scales at solid surfaces. These features, known as asperities are present down to nano-scale dimensions and result in true solid to solid contact existing only at a limited number of points accounting for only a fraction of the apparent or nominal contact area. The linearity between applied load and true contact area, arising from asperity deformation, gives rise to the linearity between static frictional force and normal force, found for typical Amonton-Coulomb type friction.

The static friction force must be overcome by an applied force before an object can move. The maximum possible friction force between two surfaces before sliding begins is the product of the coefficient of static friction and the normal force: $F_{max} = \mu_s F_n$. When there is no sliding occurring, the friction force can have any value from zero up to F_{max}. Any force smaller than F_{max} attempting to slide one surface over the other is opposed by a frictional force of equal magnitude and opposite direction. Any force larger than F_{max} overcomes the force of static friction and causes sliding to occur. The instant sliding occurs, static friction is no longer applicable—the friction between the two surfaces is then called kinetic friction.

An example of static friction is the force that prevents a car wheel from slipping as it rolls on the ground. Even though the wheel is in motion, the patch of the tire in contact with the ground is stationary relative to the ground, so it is static rather than kinetic friction.

The maximum value of static friction, when motion is impending, is sometimes referred to as limiting friction, although this term is not used universally.

Kinetic Friction

Kinetic friction, also known as dynamic friction or sliding friction, occurs when two objects are moving relative to each other and rub together (like a sled on the ground). The coefficient of kinetic friction is typically denoted as μ_k, and is usually less than the coefficient of static friction for the same materials. However, Richard Feynman comments that "with dry metals it is very hard to show any difference." The friction force between two surfaces after sliding begins is the product of the coefficient of kinetic friction and the normal force: $F_k = \mu_k F_n$.

New models are beginning to show how kinetic friction can be greater than static friction. Kinetic friction is now understood, in many cases, to be primarily caused by chemical bonding between the surfaces, rather than interlocking asperities; however, in many other cases roughness effects are dominant, for example in rubber to road friction. Surface roughness and contact area affect kinetic friction for micro- and nano-scale objects where surface area forces dominate inertial forces.

The origin of kinetic friction at nanoscale can be explained by thermodynamics. Upon sliding, new surface forms at the back of a sliding true contact, and existing surface disappears at the front of it. Since all surfaces involve the thermodynamic surface energy, work must be spent in creating the new surface, and energy is released as heat in removing the surface. Thus, a force is required to move the back of the contact, and frictional heat is released at the front.

Angle of Friction

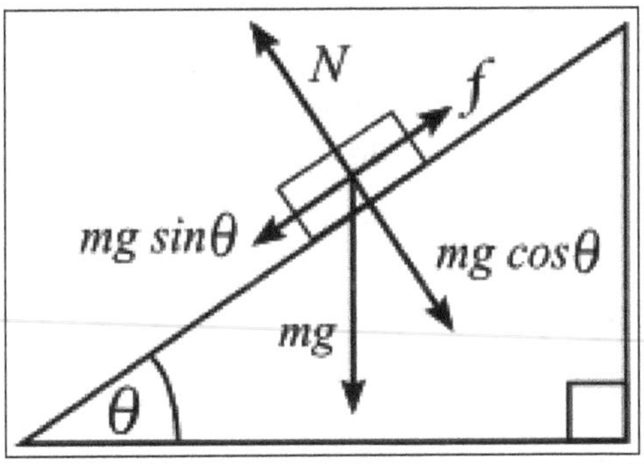

Angle of friction, θ, when block just starts to slide.

For certain applications, it is more useful to define static friction in terms of the maximum angle before which one of the items will begin sliding. This is called the angle of friction or friction angle. It is defined as:

$$\tan\theta = \mu_s$$

where θ is the angle from horizontal and μ_s is the static coefficient of friction between the objects. This formula can also be used to calculate μ_s from empirical measurements of the friction angle.

Friction at the Atomic Level

Determining the forces required to move atoms past each other is a challenge in designing nanomachines. In 2008 scientists for the first time were able to move a single atom across a surface, and measure the forces required. Using ultrahigh vacuum and nearly zero temperature (5° K), a modified atomic force microscope was used to drag a cobalt atom, and a carbon monoxide molecule, across surfaces of copper and platinum.

Limitations of the Coulomb Model

The Coulomb approximation follows from the assumptions that: Surfaces are in atomically close contact only over a small fraction of their overall area; that this contact area is proportional to the normal force (until saturation, which takes place when all area is in atomic contact); and that the frictional force is proportional to the applied normal force, independently of the contact area. The Coulomb approximation is fundamentally an empirical construct. It is a rule-of-thumb describing the approximate outcome of an extremely complicated physical interaction. The strength of the approximation is its simplicity and versatility. Though the relationship between normal force and frictional force is not exactly linear (and so the frictional force is not entirely independent of the contact area of the surfaces), the Coulomb approximation is an adequate representation of friction for the analysis of many physical systems.

When the surfaces are conjoined, Coulomb friction becomes a very poor approximation (for example, adhesive tape resists sliding even when there is no normal force, or a negative normal force). In this case, the frictional force may depend strongly on the area of contact. Some drag racing tires are adhesive for this reason. However, despite the complexity of the fundamental physics behind friction, the relationships are accurate enough to be useful in many applications.

Negative Coefficient of Friction

As of 2012, a single study has demonstrated the potential for an effectively negative coefficient of friction in the low-load regime, meaning that a decrease in normal force leads to an increase in friction. This contradicts everyday experience in which an increase in normal force leads to an increase in friction. This was reported in the journal Nature in October 2012 and involved the friction encountered by an atomic force microscope stylus when dragged across a graphene sheet in the presence of graphene-adsorbed oxygen.

Numerical Simulation of the Coulomb Model

Despite being a simplified model of friction, the Coulomb model is useful in many numerical simulation applications such as multibody systems and granular material. Even its most simple expression encapsulates the fundamental effects of sticking and sliding which are required in many applied cases, although specific algorithms have to be designed in order to efficiently numerically integrate mechanical systems with Coulomb friction and bilateral or unilateral contact. Some quite nonlinear effects, such as the so-called Painlevé paradoxes, may be encountered with Coulomb friction.

Dry Friction and Instabilities

Dry friction can induce several types of instabilities in mechanical systems which display a stable behaviour in the absence of friction. These instabilities may be caused by the decrease of the friction force with an increasing velocity of sliding, by material expansion due to heat generation during friction (the thermo-elastic instabilities), or by pure dynamic effects of sliding of two elastic materials (the Adams-Martins instabilities). The latter were originally discovered in 1995 by George G. Adams and João Arménio Correia Martins for smooth surfaces and were later found in periodic rough surfaces. In particular, friction-related dynamical instabilities are thought to be responsible for brake squeal and the 'song' of a glass harp, phenomena which involve stick and slip, modelled as a drop of friction coefficient with velocity.

A practically important case is the self-oscillation of the strings of bowed instruments such as the violin, cello, hurdy-gurdy, erhu, etc.

Frictional instabilities can lead to the formation of new self-organized patterns (or "secondary structures") at the sliding interface, such as in-situ formed tribofilms which are utilized for the reduction of friction and wear in so-called self-lubricating materials.

Laws of Dry Friction

The elementary property of sliding (kinetic) friction were discovered by experiment in the 15th to 18th centuries and were expressed as three empirical laws:

- Amontons' First Law: The force of friction is directly proportional to the applied load.
- Amontons' Second Law: The force of friction is independent of the apparent area of contact.
- Coulomb's Law of Friction: Kinetic friction is independent of the sliding velocity.

Wear

Rear (driven) bicycle sprockets. New, left, shows no wear. Right, used, shows obvious wear from being driven clockwise.

Wear is the damaging, gradual removal or deformation of material at solid surfaces. Causes of wear can be mechanical (e.g., erosion) or chemical (e.g., corrosion). The study of wear and related processes is referred to as tribology.

Wear in machine elements, together with other processes such as fatigue and creep, causes functional surfaces to degrade, eventually leading to material failure or loss of functionality. Thus, wear has large economic relevance as first outlined in the Jost Report.

Wear of metals occurs by plastic displacement of surface and near-surface material and by detachment of particles that form wear debris. The particle size may vary from millimeters to nanometers. This process may occur by contact with other metals, nonmetallic solids, flowing liquids, solid particles or liquid droplets entrained in flowing gasses.

The wear rate is affected by factors such as type of loading (e.g., impact, static, dynamic), type of motion (e.g., sliding, rolling), temperature, and lubrication, in particular by the process of deposition and wearing out of the boundary lubrication layer. Depending on the tribosystem, different wear types and wear mechanisms can be observed.

Classification of Wear

Wear is commonly classified according to so-called wear types, which occur in isolation or complex interaction. Common types of wear include:

- Adhesive wear.
- Abrasive wear.
- Surface fatigue.
- Fretting wear.

- Erosive wear.
- Corrosion and oxidation wear.

Other, less common types of wear are impact, cavitation and diffusive wear.

Each wear type is caused by one or more wear mechanisms. For example, the primary wear mechanism of adhesive wear is adhesion. Wear mechanisms and/or sub-mechanisms frequently overlap and occur in a synergistic manner, producing a greater rate of wear than the sum of the individual wear mechanisms.

Adhesive Wear

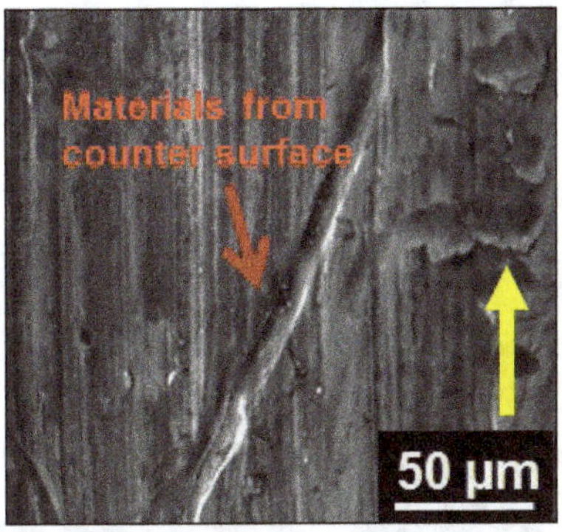

SEM micrograph of adhesive wear (transferred materials) on 52100 steel sample sliding against Al alloy. (Yellow arrow indicate sliding direction).

Adhesive wear can be found between surfaces during frictional contact and generally refers to unwanted displacement and attachment of wear debris and material compounds from one surface to another. Two adhesive wear types can be distinguished:

- Adhesive wear is caused by relative motion, "direct contact" and plastic deformation which create wear debris and material transfer from one surface to another.
- Cohesive adhesive forces, holds two surfaces together even though they are separated by a measurable distance, with or without any actual transfer of material.

Generally, adhesive wear occurs when two bodies slide over or are pressed into each other, which promote material transfer. This can be described as plastic deformation of very small fragments within the surface layers. The asperities or microscopic high points (surface roughness) found on each surface affect the severity of how fragments of oxides are pulled off and added to the other surface, partly due to strong adhesive forces between atoms, but also due to accumulation of energy in the plastic zone between the asperities during relative motion.

The type of mechanism and the amplitude of surface attraction varies between different materials but are amplified by an increase in the density of "surface energy". Most solids will adhere on contact to some extent. However, oxidation films, lubricants and contaminants naturally occurring generally suppress adhesion, and spontaneous exothermic chemical reactions between surfaces generally produce a substance with low energy status in the absorbed species.

Adhesive wear can lead to an increase in roughness and the creation of protrusions (i.e., lumps) above the original surface. In industrial manufacturing, this is referred to as galling, which eventually breaches the oxidized surface layer and connects to the underlying bulk material, enhancing the possibility for a stronger adhesion and plastic flow around the lump.

Abrasive Wear

Abrasive wear occurs when a hard rough surface slides across a softer surface. ASTM International defines it as the loss of material due to hard particles or hard protuberances that are forced against and move along a solid surface.

Deep 'groove' like surface indicates abrasive wear over cast iron (yellow arrow indicate sliding direction).

Abrasive wear is commonly classified according to the type of contact and the contact environment. The type of contact determines the mode of abrasive wear. The two modes of abrasive wear are known as two-body and three-body abrasive wear. Two-body wear occurs when the grits or hard particles remove material from the opposite surface. The common analogy is that of material being removed or displaced by a cutting or plowing operation. Three-body wear occurs when the particles are not constrained, and are free to roll and slide down a surface. The contact environment determines whether the wear is classified as open or closed. An open contact environment occurs when the surfaces are sufficiently displaced to be independent of one another.

There are a number of factors which influence abrasive wear and hence the manner of material removal. Several different mechanisms have been proposed to describe the manner in which the material is removed. Three commonly identified mechanisms of abrasive wear are:

- Plowing,
- Cutting,
- Fragmentation.

Plowing occurs when material is displaced to the side, away from the wear particles, resulting in the formation of grooves that do not involve direct material removal. The displaced material forms ridges adjacent to grooves, which may be removed by subsequent passage of abrasive particles.

Cutting occurs when material is separated from the surface in the form of primary debris, or microchips, with little or no material displaced to the sides of the grooves. This mechanism closely resembles conventional machining.

Fragmentation occurs when material is separated from a surface by a cutting process and the indenting abrasive causes localized fracture of the wear material. These cracks then freely propagate locally around the wear groove, resulting in additional material removal by spalling.

Abrasive wear can be measured as loss of mass by the Taber Abrasion Test according to ISO 9352 or ASTM D 4060.

Surface Fatigue

Surface fatigue is a process in which the surface of a material is weakened by cyclic loading, which is one type of general material fatigue. Fatigue wear is produced when the wear particles are detached by cyclic crack growth of microcracks on the surface. These microcracks are either superficial cracks or subsurface cracks.

Fretting Wear

Fretting wear is the repeated cyclical rubbing between two surfaces. Over a period of time fretting which will remove material from one or both surfaces in contact. It occurs typically in bearings, although most bearings have their surfaces hardened to resist the problem. Another problem occurs when cracks in either surface are created, known as fretting fatigue. It is the more serious of the two phenomena because it can lead to catastrophic failure of the bearing. An associated problem occurs when the small particles removed by wear are oxidised in air. The oxides are usually harder than the underlying metal, so wear accelerates as the harder particles abrade the metal surfaces further. Fretting corrosion acts in the same way, especially when water is present. Unprotected bearings on large structures like bridges can suffer serious degradation in behaviour, especially when salt is used the during winter to deice the highways carried by the bridges. The problem of fretting corrosion was involved in the Silver Bridge tragedy and the Mianus River Bridge accident.

Erosive Wear

Erosive wear can be defined as an extremely short sliding motion and is executed within a short time interval. Erosive wear is caused by the impact of particles of solid or liquid against the surface of an object. The impacting particles gradually remove material from the surface through repeated deformations and cutting actions. It is a widely encountered mechanism in industry. Due to the nature of the conveying process, piping systems are prone to wear when abrasive particles have to be transported.

The rate of erosive wear is dependent upon a number of factors. The material characteristics of the particles, such as their shape, hardness, impact velocity and impingement angle are primary factors along with the properties of the surface being eroded. The impingement angle is one of the most important factors and is widely recognized in literature. For ductile materials, the maximum wear rate is found when the impingement angle is approximately 30°, whilst for non-ductile materials the maximum wear rate occurs when the impingement angle is normal to the surface.

Corrosion and Oxidation Wear

Corrosion and oxidation wear occurs both in lubricated and unlubricated contacts. The fundamental causes are chemical reactions between the worn material and the corroding medium. Wear caused by a synergistic action of tribological stresses and corrosion is also called tribocorrosion.

Theories of Wear

Adhesive Wear Theory

Wear is due to welded asperities breaking off from the surface. Material hardness (higher) is important but the Hardness/Elasticity ratio is more significant because it represents the yield strain of a material. Both the bulk and surface hardness are important here, so hardening techniques are often used. To reduce wear use dissimilar materials which are metallurgically incompatible and put a soft layer on the surface to protect the surface until it has run in.

In adhesive wear, wear is proportional to load and independent of velocity and apparent area of contact. The wear coefficient K would be 1 if all asperities contacts led to detachment, but K is generally 10^-6 or 10^-7, so very few asperity contacts result in a wear particle. However, frictional heating can be important and is a function of velocity.

Abrasive Theory

This is caused by sharp particles indenting a surface. They generally cause a furrow which displaces material on either side of the furrow. Only a certain portion of abrasives (generally 10%) have a good angle of attack for cutting and cause wear. Abrasive wear takes place when with a soft surface and hard counterface. More important than the regular roughness value is the sharpness of asperities (sharper asperities are worse).

Third body wear wear is common and falls into this category. The first subcategory is adventitious which is caused by external material and a filter or seal can be used to prevent it. The second comes from wear particles and is dependent on the properties of the oxide of the base material.

Fatigue Wear Theory

There is macro-fatigue wear (also referred to as rolling contact fatigue and pitting) as well as micro-fatigue wear. The presence of macro-fatigue wear means that a part has failed. As stress increases, life decreases. Pit depth will be a function of the depth of the maximum shear stress in the contact.

Corrosion Wear Theory

This is caused by corrosion of the surface. It is more common when additives are used to prevent adhesive wear because the additives create a surface layer which is actually a corrosive degradation of the material. The corrosive wear process begins with a corrosion film growing on the surface. It increases in thickness until the film reaches a critical wear depth and then breaks off. The process begins again. The growth rate is governed by thermal processes.

Lubricated Wear

This type of wear is governed by the ratio of the film thickness to roughness which is referred to as the lambda ratio and is the inverse of the D ratio. Conformal contacts are little problem because they have a high lambda ratio. However, they can be troublesome if the shaft is off center, if material enters the bearing and bridges the lubricant film, or if there is no tapered surface.

In EHL there are two important regions. The inlet region which determines the film thickness and the converged region where there is a high pressure. Some EHL contacts (soft EHL) are slightly different because materials have a low modulus so viscosity is not affected. Gear failures are in this regime and occur due to rolling contact fatigue and scuffing.

Wear Prevention

Wear may be prevented in many different ways. The most common techniques is to use wear resistant materials or simply by changing the material properties, e.g. by hardening. Wear can also be prevented by keeping the surfaces apart from each other. Oil and grease are commonly used for this purpose.

A third method is to design for reduced wear. For instance a pipe can be given a large diameter that reduces the water velocity and a pump can be designed to avoid entrapped particles and reduce wear.

Grindex innovation of a new patented suction cover reduces wear.

Wear Testing and Wear Measurement

Wear test is carried out to predict the wear performance and to investigate the wear mechanism.

Two specific reasons are as follows:

- From a material point of view, the test is performed to evaluate the wear property of a material so as to determine whether the material is adequate for a specific wear application.
- From a surface engineering point of view, wear test is carried out to evaluate the potential of using a certain surface engineering technology to reduce wear for a specific application, and to investigate the effect of treatment conditions (processing parameters) on the wear performance, so that optimised surface treatment conditions can be realised.

Wear test is performed in three levels, namely:

- Laboratory Test.

- Component Simulation Test.
- In-service test.

We may use an example to describe the difference among each type of test. A new surface engineering (SE) technology has been developed, which could be potentially used to improve the wear resistance of parts for a metal-on-metal hip joint (for human body). Perhaps the ideal and logical sequence of wear testing in this example will be as follows:

- Stage 1: Laboratory test with small samples are initially carried out under testing conditions simulated insofar as conveniently possible to determine whether the surface engineering technology warrants further consideration, and if so, to find out under what treatment conditions, the highest wear resistance improvement can be achieved.
- Stage 2: In the next stage, the optimised surface treatment condition will be used to treat some real joint parts, e.g. femoral head or cup, which will then be tested in a hip-joint simulator, with the testing conditions being controlled as close as possible to those for a real human joint, for example, lubricated with body liquid, temperature around 37 °C, and moving like a human walking or running.
- Stage 3: Only when the surface treated parts have survived the simulation tests, and indeed show considerably improved wear resistance yet without losing other properties (e.g. corrosion and bio-compatibility), can the in-service tests be carried out. The surface treated joint parts will be implanted in a human body and tested (monitored) for a pro-longed period of time if there is no immediate side effect after implantation.

As we can see, simulation test and in-service test produce more reliable results, however, laboratory test is cheap, safe and quick and thus serve a most useful function. In the following sections, we will introduce the methods for laboratory wear test. The working principles of three widely used laboratory wear testers will be given, followed by an introduction of wear measurement.

Laboratory Wear Testing Method

An apparatus for wear testing is termed as wear tester, tribotester or tribometer. The prefix of "tribo-" refers to wear, friction and lubrication. Many, probably more than several hundreds, different wear testing arrangements and procedures are used in laboratories around the world. However big difference of one arrangement as compared with another, a wear tester will always involve two components loaded against and relatively moving each other. The movement can be driven by a motor or by an electro-magnetic device. For convenient purpose, the material or component being investigated is normally referred to as specimen, the other termed as counterface. As examples, the working principle of three most widely used laboratory wear testers are given below.

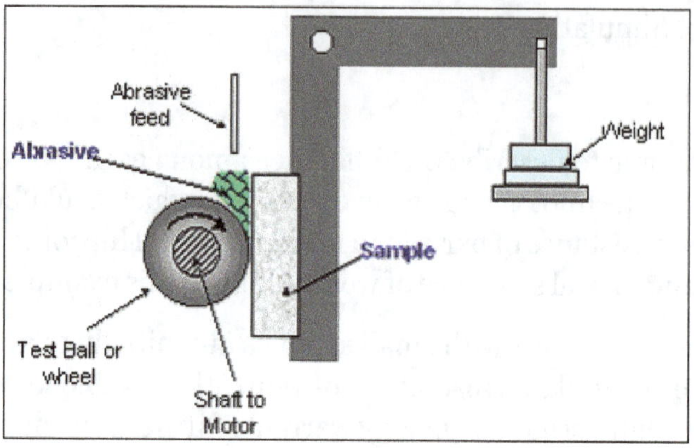

An abrasive wear tester.

Example of an Abrasive Wear Tester

Figure is a schematic of an abrasive wear tester, in which a wheel or a ball is driven by a motor, rotating and sliding against a fixed sample in the presence of abrasive particles. The specimen is in the form of a plate or a block. Contact pressure is controlled by dead weight through a loading lever. The abrasive particles, such as silica, are added through a nozzle connecting to a hopper above, giving a three-body wear situation. After a set time of running, the sample is removed, and wear loss is measured. The parameters to be controlled include contact load, sliding speed, type of abrasive particles and its flow rate.

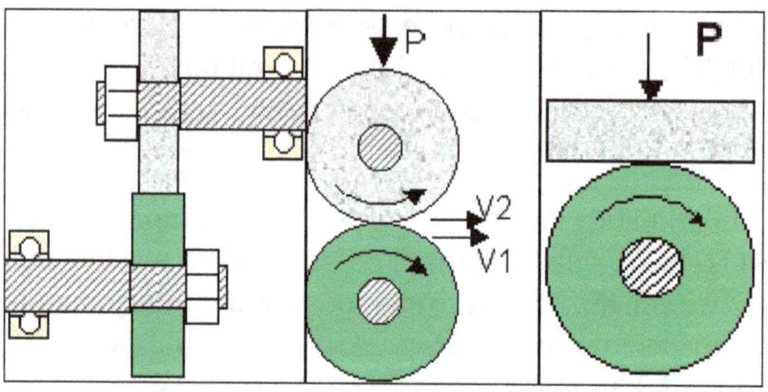

Sample arrangement in a rolling sliding wear test.

Example of Rolling Sliding Wear Tester

Rolling-sliding wear tester is the most popular tribometer for investigating wear as well as frictional behaviour of a materials under conditions of rolling, sliding, or a combination of both. Two discs (wheels), as show in figure), are fixed to two parallel shafts and pressed against each other under a constant contact load. Driven by a motor through a train of gear, the specimens are rotating along with the shafts. The rotating speed can be controlled, so that when the linear speeds of two wheels are equal at the contact point (V1=V2), a pure rolling contact is achieved. When V1 and V2 are different (V1? V2) and both wheels are rotating, a combined rolling-sliding can be realised. Whilst when one of the specimen is fixed, and the other is rotating, then wear is a pure sliding. In this case, the fixed specimen can be a block, so that a name of block-on-wheel is used. Abrasive particles may be added to the contact area, achieving a three body abrasive wear testing.

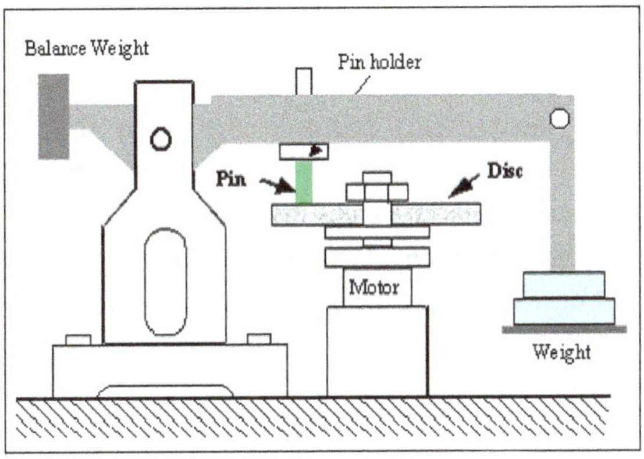

Pin-on-disc-machine.

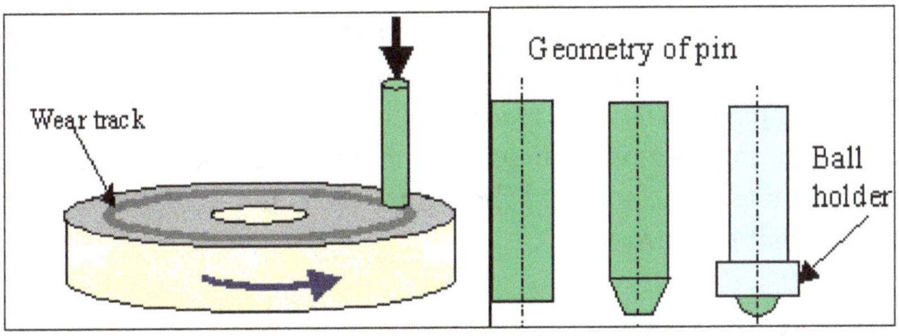

Schematic of a pin-on-disc wear test and the arrangement of samples.
Arrangement on samples, geometry of pin.

Example of Pin-on-disc Wear Tester

In a pin-on-disc wear tester, a pin is loaded against a flat rotating disc specimen such that a circular wear path is described by the machine. The machine can be used to evaluate wear and friction properties of materials under pure sliding conditions. Either disc or pin can serve as specimen, while the other as counterface. Pin with various geometry can be used. A convenient way is to use ball of commercially available materials such as bearing steel, tungsten carbide or alumina (Al_2O_3) as counterface, so that the name of ball-on-disc is used.

Wear Measurement/Quantification

Wear measurement is carried out to determine the amount of materials removed (or worn away) after a wear test, (and in reality after a part in service for a period of time). The material worn away can be expressed either as weight (mass) loss, volume loss, or linear dimension change depending on the purpose of the test, the type of wear, the geometry and size of the test specimens, and sometimes on the availability of a measurement facility.

Common techniques of wear measurement include using a precision balance to measure the weight (mass) loss, profiling surfaces, or using a microscope to measure the wear depth or cross-sectional area of a wear track so as to determine the wear volume loss or linear dimensional change.

Mass loss

Mass loss measurement by a precision balance is a convenient method for wear measurement, especially when the worn surface is irregular and unsymmetrical in shape. Sample to be measured is carefully cleaned, and the weight is measured before and after a wear test. The difference in weight before and after test represents the weight loss caused by wear. The unit can be gram (g) or milligram (μg).

Volume Loss

Wear volume is normally calculated from the wear track (scar) depth, length, width and/or scar profile according to the geometry of the wear track/scar. A surface profilometer, e.g. a stylus type, or sometimes a microscope with scale is used for the measurement. The reporting unit of wear volume loss is mm3 or μm3. Wear volume loss enables a better comparison of wear among materials having different densities. However, it is not easy to measure volume loss when a wear track is irregular. In this case, mass loss may be measured first, and the volume loss is calculated if the materials is uniform and its density is known.

Linear Dimension Loss

Measuring wear by linear dimension change is very useful in many engineering situations, where certain dimension such as length, thickness or diameter is more critical to the normal function of the system. A surface profilometer, e.g. a stylus type, a micrometer or a microscope can be used. The unit for linear dimension loss can be μm or mm.

Wear Rate

Wear rates are calculated results reflecting wear mass loss, volume loss or linear dimension change under unit applied normal force and/or unit sliding distance. Wear rate can be expressed in many different ways.

Wear Resistance

Wear resistance is a term frequently used to describe the anti-wear properties of a material. However, the scientific meaning of wear resistance is vague, and there is no specific unit to describe wear resistance. Nevertheless, the inverse of mass loss or volume loss is sometimes used as the (relative) wear resistance. The ratio of wear loss for a reference material over that of the investigated material under same testing conditions can also be used as relative wear resistance. In any case, if a numeric value of wear resistance is given, its meaning should be clearly indicated.

Table: Wear measurement methods and typical units for wear quantification.

	Measurement Methods	Units of wear	Units of wear rate
Mass loss	Direct measurement by a precision balance. Calculated from volume loss for known density material.	ig	μ.g/m, g/m, μg/N, g/N, μgl(N m), g/(N m).
Volume loss	Calculated from depth, width, wear profile and/or other dimensions data of a wear track. Surface or microscopy techniques can be used for the measurement. Calculated from mass loss for known density material.	mm3	mm3/m mm3!N mm3/(N m),
Linear dimension	Direct measurement by surface microscopy profilometry and other dimension measurement techniques.	μm mm	μmlyear mm/ year

References

- Friction: tribonet.org, Retrieved 28 July, 2020
- Bigoni, D.; Noselli, G. (2011). "Experimental evidence of flutter and divergence instabilities induced by dry friction". Journal of the Mechanics and Physics of Solids. 59 (10): 2208–2226. Bibcode:2011JMPSo..59.2208B. CiteSeerX 10.1.1.700.5291. doi:10.1016/j.jmps.2011.05.007
- Wear-theory-and-mechanisms: tribos.wordpress.com, Retrieved 26 June, 2020
- Wear-testing-measurement: emrtk.uni-miskolc.hu, Retrieved 15 May, 2020
- Chattopadhyay, R. (2001). Surface Wear - Analysis, Treatment, and Prevention. OH, USA: ASM-International. ISBN 978-0-87170-702-4

Lubrication and Lubricants

3

- Lubrication
- Automatic Lubricator
- Basic Modes of Lubrication
- Boundary Lubrication
- Lubrication Theory
- Theory of Hydrodynamic Lubrication
- Reynold's Equations and its Limitations
- Viscosity Index
- Lubricants
- Types of Lubricants
- Lubrication Mechanism
- Properties of Lubricants
- Lubricant Additives
- Lubricant Selection
- Hydrostatic Lubrication
- Elasto- Hydrodynamic Lubrication

Lubrication can be defined as the process in which lubricants are used for reducing friction and wear between two surfaces. A few types of lubricants include mineral and synthetic oils, biolubricants, solid lubricants, etc. This chapter closely examines lubrication and lubricants to provide an extensive understanding of the subject.

Lubrication

The primary purpose of lubrication is to reduce wear and heat between contacting surfaces in relative motion. While wear and heat cannot be completely eliminated, they can be reduced to negligible or acceptable levels. Because heat and wear are associated with friction, both effects can be minimized by reducing the coefficient of friction between the contacting surfaces. Lubrication is also used to reduce oxidation and prevent rust; to provide insulation in transformer applications; to transmit mechanical power in hydraulic fluid power applications; and to seal against dust, dirt, and water.

Lubrication Regimes

The modern period of lubrication began with the work of Osborne Reynolds. Reynold's research was concerned with shafts rotating in bearings and cases this show in figure. When a lubricant was applied to the shaft, Reynolds found that a rotating shaft pulled a converging wedge of lubricant between the shaft and the bearing. He also noted that as the shaft gained velocity, the liquid flowed between the two surfaces at a greater rate. This, because the lubricant is viscous, produces a liquid pressure in the lubricant wedge that is sufficient to keep the two surfaces separated. Under ideal conditions, Reynolds showed that this liquid pressure was great enough to prevent direct contact between the metal surfaces. Figure taking a plain journal bearing as example, figure which is known as Stribeck curve summarizes the lubrication regimes by describing the relationship between speed, load, oil viscosity, oil film thickness, and friction.

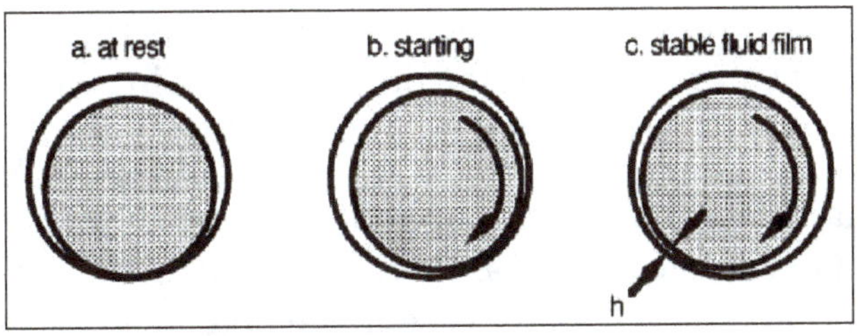

Three positions of shaft in a bearing.

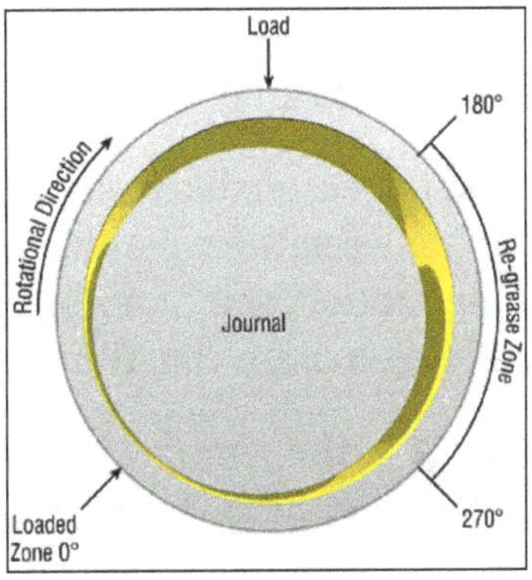

Plain Journal bearing.

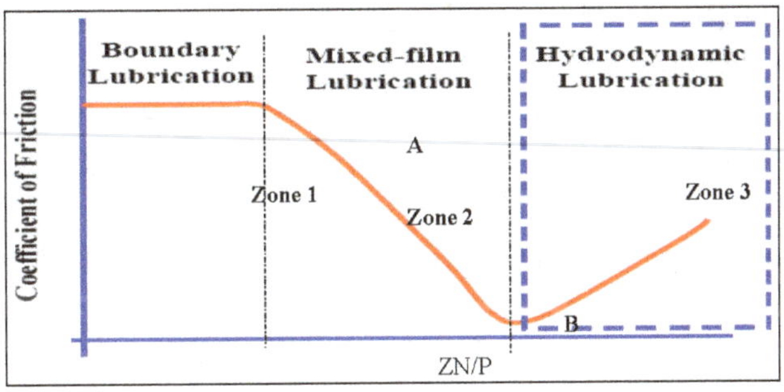

Stribeck curve.

In this graph, the coefficient of friction is plotted against the expression ZN/P (sometimes referred to as the Hersey number).

$$\text{Where } ZN/P = \frac{\text{oil viscosity} \times \text{shaft speed}}{\text{bearing pressure}}$$

As shown there are three distinct zones separated by points A and B. At B the oil film is just thick enough to ensure that there is no contact between asperities on the shaft and bearing surfaces. Smoother surfaces shift B to the left, while at point A the oil film thickness reduces virtually to nil. Zone 2, between A and B is known as the zone of mixed lubrication. Mixed-film lubrication is unstable at which increase in lubrication temperature causes further increases in lubrication temperature.

Mixed-film Lubrication

Hydrodynamic Lubrication

Basically, lubrication is governed by one of two principles: Hydrodynamic lubrication and boundary lubrication. In the former, a continuous full-fluid film separates the sliding surfaces. In the latter, the oil film is not sufficient to prevent metal-to-metal contact. Hydrodynamic lubrication is the more common, and it is applicable to nearly all types of continuous sliding action where extreme pressures are not involved. Whether the sliding occurs on flat surfaces, as it does in most thrust bearings, or whether the surfaces are cylindrical, as in the case of journal (plain or sleeve) bearings, the principle is essentially the same.

It would be reasonable to suppose that, when one part slides on another, the protective oil film between them would be scraped away. Except under some conditions of reciprocating motion, this is not necessarily true at all. With the proper design, in fact, this very sliding motion constitutes the means of creating and maintaining that film.

In zone 3 is the zone of hydrodynamic or fluid film lubrication where there is no wear because there is no contact between the surfaces. Hydrodynamic Lubrication is often referred to as stable lubrication. There are four essential elements in hydrodynamic lubrication, a liquid, relative motion, the viscous properties of the liquid, and the geometry of the surfaces between which the convergent wedge of fluid is produced. Only friction present in a hydrodynamic lubrication system is the friction of the lubricant itself, it would make sense to have a less viscous fluid in order to minimize friction: the less viscous a liquid the lower the friction. Too low of a viscosity jeopardizes our system though. We have to be very careful that the distance between the two surfaces is greater than the largest surface defect. The distance between the two surfaces decreases with higher loads on the bearing, less viscous fluids, and lower speeds. The surface geometry is also very important. The surfaces have to be such that a converging wedge of fluid can develop between the surfaces, allowing the hydrodynamic pressure of the lubricant to support the load of the shaft or moving surface. Hydrodynamic lubrication is an excellent method of lubrication since it is possible to achieve coefficients of friction as low as 0.001, and there is no wear between the moving parts. Special attention must be paid to the heating of the lubricant by the frictional force since viscosity is temperature dependent. One method of accomplishing this is to cycle the lubricant through a cooling reservoir in order to maintain the desired viscosity of the fluid. Another way of handling the heat dissipation is to use commercially available additives to decrease the viscosity's temperature dependence which are known as viscosity index improvers.

The formation of fluid film is influenced by the following factors:

- The contact surfaces must meet at a slight angle to allow formation of the lubricant wedge.
- The fluid viscosity must be high to maintain adequate film thickness to separate the contacting surfaces at operating speeds.
- The fluid must be adhering to the contact surfaces for conveyance into the pressure area to support the load.
- The fluid must be distributing itself completely within the bearing clearance area.
- The operating speed must be sufficient to allow formation and maintenance of the fluid film.
- The contact surfaces of bearings and journals must be smooth and free from sharp surfaces that will disrupt the fluid film.

Boundary Lubrication

The oil film has become so thin in Zone 1 that there is no hydrodynamic contribution and only boundary lubrication which is defined by Campbell in 1969 as the lubrication by a liquid under conditions where the solid surfaces are so close together that appreciable contact between opposing asperities is possible. The friction and wear in boundary lubrication are determined predominantly by interaction between the solids and between the solids and the liquid. The bulk flow properties of the liquid play little or no part in the friction and wear behavior.

Boundary lubrication is effective when a complete fluid film does not develop between potentially rubbing surfaces, the film thickness may be reduced to permit momentary dry contact between wear surface high points or asperities. Boundary lubrication occurs whenever any of the essential factors that influence formation of a full fluid film are missing. The most common example of boundary lubrication includes bearings, which normally operate with fluid film lubrication but experience boundary lubricating conditions during routine starting and stopping of equipment. Other examples include gear tooth contacts and reciprocating equipment.

Generally, the best additives are active organic compounds with long chain molecules and active end groups. These compounds bind tightly and intricately with each other, forming a film that builds up on the surface of the metal itself. This results in a thin film that is very difficult to penetrate. When two surfaces, each covered with a boundary layer, come in contact with each other they tend to slide along their outermost surfaces, with the actual faces of the surfaces rarely making contact with each other. Liquids are rarely good boundary lubricants. The best boundary lubricants are solids with long chains of high inter-chain attraction, low shear resistance so as to slip easily, and a high temperature tolerance. The boundary lubricant should also, obviously, be able to maintain a strong attachment to the surfaces under high temperatures and load pressures.

The most common boundary lubricants are probably greases. Greases are so widely used because they have the most desirable properties of a boundary lubricant. They not only shear easily, they flow. They also dissipate heat easily; form a protective barrier for the surfaces, preventing dust, dirt, and corrosive agents from harming the surfaces Scheme.

Base Stock

Petroleum is one of the naturally occurring hydrocarbons that frequently include natural gas, natural bitumen, and natural wax. According to the most generally accepted theory today, petroleum was formed by the decomposition of organic refuse, aided by high temperatures and pressures, over a vast period of geological time.

Although petroleum occurs, as its name indicates, among rocks in the earth, it sometimes seeps to the surface through fissures or is exposed by erosion. The existence of petroleum was known to primitive man, since surface seepage, often sticky and thick, was obvious to anyone passing by prehistoric animals were sometimes mired in it, but few human bones have been recovered from these tar pits. Early man evidently knew enough about the danger of surface seepage to avoid it.

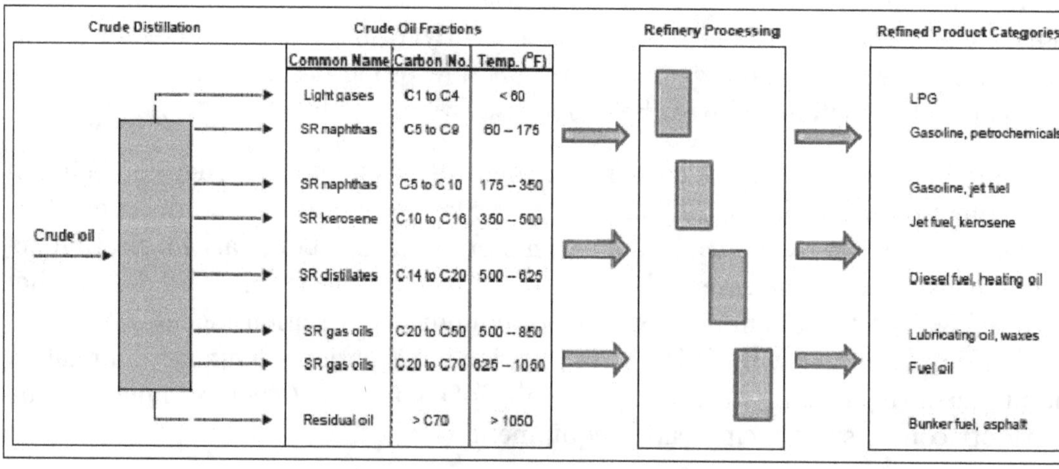

Schematic View of Crude Oil Distillation.

The petroleum remaining from the distillation is thick like pitch; if the distillation has been pushed far, the residuum will flow only languidly in the retort, and in cold weather it becomes a soft solid, resembling much the maltha or mineral pitch figure shows that the distillation of crude oil.

Base stocks are refined from crude oil to obtain products with the best lubricating properties. Base stocks generally make up 80-95% of a typical engine oil and 5% additives. Base stock is used to describe plain mineral oil. The physical properties of an oil depend on its base stock. In most cases it is chemically inert there are three sources of base stock: biological, mineral and synthetic. The oils manufactured from these sources exhibit different properties and they are suitable for different applications. For example:

- Biological oils are suitable in applications where the risk of contamination must be reduced to a minimum, for example, in the food or pharmaceutical industry. They are usually applied to lubricate kilns, bakery ovens, etc. There can be two sources of this type of oil: Vegetable and animal. Examples of vegetable oils are: Castor, palm and rape-seed oils, while the examples of animal oils are: Sperm, fish and wool oils from sheep (lanolin).

- Mineral oils are the most commonly used lubricants throughout industry. They are petroleum based and are used in applications where temperature requirements are moderate. Typical applications of mineral oils are to gears, bearings, engines, turbines, etc.

- Synthetic oils are artificially developed substitutes for mineral oils. They are specifically developed to provide lubricants with superior properties to mineral oils. For example, temperature resistant synthetic oils are used in high performance machinery operating at high temperatures. Synthetic oils for very low temperature applications are also available.

Applications of Lubrication

Proper lubrication is a key factor affecting the performance and longevity of rotating mechanical equipment. From temperatures of over 1,000 degrees Fahrenheit to speeds in the tens of thousands of revolutions per minute, the moving parts of industrial machinery experience colossal amounts of stress as part of their everyday operation. One of the main factors that allow equipment to operate under such extreme circumstances is the lubrication used to reduce friction between moving surfaces. Understanding the types, selection, and applications of the lubrication found in an industrial facility is vital knowledge for operators and technicians.

Oils

Liquid lubricants, generally referred to as oils or lube oils, share the properties of all liquids, are able to flow, and take the shape of their containers. The properties of lubricating oils make them suitable for a wide range of applications.

- Remove Heat: Since oil can flow freely, it has the ability to absorb and carry away the heat generated by the components it flows over.
- Cleaning: As oil flows through a system, it will pick up and suspend contaminants. The oil can then be passed through filters or settling areas that allow the contaminants to be removed from the system.
- Level Control: Using sight glasses or level indicators, the exact amount of oil in a system can be determined and controlled.
- Replaceability: As oil is flowing through a system, it can be removed and new oil added without interrupting component operation. This can save costs on both maintenance hours and equipment downtime.
- Representative Sampling: Oil gets mixed as it moves through a system, allowing a fairly representative sample to be taken and analyzed for lubricant and component health.
- Leakage: Although it may not seem like a good thing, oil leaking from worn seals and gaskets gives a positive indication of components that need to be replaced. This can prevent much more serious and expensive damage from occurring.

There are several types of lubricating oils used in industrial applications, each with its own strengths and weaknesses:

- Mineral Oils: Mineral oils are composed of a complex mixture of hydrocarbons and are produced by processing crude, or petroleum, oil. One of the main advantages of using mineral oils is that they are manufactured on a large-scale basis during the refinement of petroleum and are, therefore, much cheaper than the alternatives.
- Synthetic Oils: Synthetic oils, on the other hand, are artificially made typically by combining organic chemicals into specific mixtures. Advantages of synthetic oils include better viscosity at high and low temperatures, better chemical

stability, better resistance to oxidation, and good fire resistance. However, synthetics are more expensive than mineral oils and susceptible to decomposition in certain chemical environments.

- Fixed Oils: Fixed oils, or biolubricants, are processed from animal or vegetable sources. They are rarely used by themselves as they tend to break down at high temperatures. Instead, they are added to mineral oils and greases.

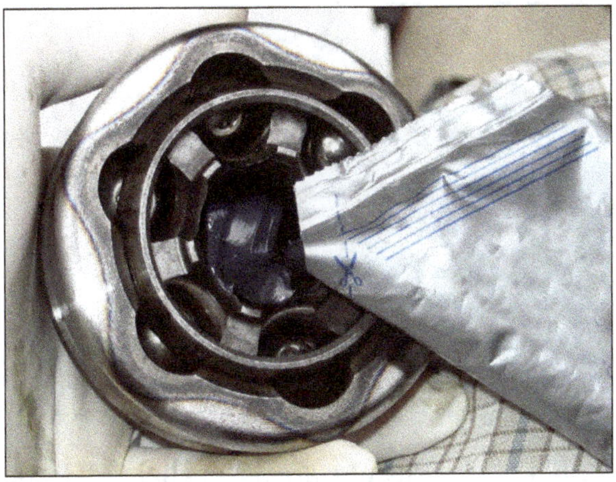

Greases

Oils are not always the best choice when it comes to lubricating components. Some situations require the lubricant to stick to a component, while others require lubrication for components that are typically not accessible. Semi-solid lubricants, commonly referred to as greases, share many of the same properties of their liquid counterparts, but are designed to stay in one place or stick to the parts they are meant to lubricate. The following are some advantages of using grease to lubricate machinery:

- Application: Grease will stick to the surface to which it is applied; even vertical surfaces open to the environment. This allows grease to be used in applications where oil would be ineffective.

- Independent of System: Unlike oil, grease does not require a conditioning system or pumps for transportation.

- Startup: Grease will stick to the component it is meant to lubricate, unlike oil, which will drain away when a machine is shut down. These results in the machine having lubrication the instant it starts up, instead of waiting on an oil supply.

- No Leakage: Even if a machine has worn or damaged gaskets or seals, grease will still stick to the components and not leak out.

- Sealant: Extra grease can be packed around the components and seals to act as a barrier from contaminants and water. Also, adding more grease to a chamber already filled with it will force out the old grease, taking any suspended contaminants with it.

- Solid Additives: Solid lubricants like graphite can be added to grease to help provide lubrication. If these were added to an oil system, the solids would be carried away by the oil and eventually filtered out.

Dry Lubricants

Solid lubricants, also called dry lubricants, are used where extreme pressures, temperatures, or operating conditions, such as vacuum, dust or dirt, and corrosive contaminants prevent using mineral oils and greases. The most common solid lubricants are graphite and molybdenum disulfide. There are two ways solid lubricants are typically applied to equipment:

- They can be added to greases and oils, which carry the solids to the desired surfaces. If the grease or oil is squeezed out from the surface by extreme pressure, the solids will remain behind.
- They can be sprayed on the surface to be lubricated as part of a mixture. The solid will be mixed with a volatile chemical, like alcohol, which will evaporate and leave a layer of solid lubricant.

Lubrication is a vital part of industrial machinery. Understanding the different types and their correct applications will help ensure that your equipment maintains peak performance and reliable operation for years to come.

Automatic Lubricator

An automatic lubricator, is a device fitted to a steam engine to supply lubricating oil to the cylinders and, sometimes, the bearings and axle box mountings as well. There are various types of automatic lubricator, which include various designs of displacement, hydrostatic and mechanical lubricators.

Modern locomotive lubricators: The centre item is a mechanical lubricator for the cylinders, operated by the connecting lever seen below it (or by the hand wheel, for priming). The smaller one to the right is a drip-feed lubricator.

Displacement Lubricator

Ramsbottom Type

The displacement lubricator was introduced in the United Kingdom in 1860 by John Ramsbottom. It operates by allowing steam to enter a closed vessel containing oil. After condensing, the water sinks to the bottom of the vessel, causing the oil to rise and overflow into delivery pipes. The oil from the delivery pipes is introduced into the steam pipe, where it is atomised and carried to the valves and cylinders.

In early applications in steam locomotives, either two displacement lubricators (one for each cylinder) would be positioned at the front of the boiler near the valves, often on either side of the smokebox or one lubricator would be placed behind the smokebox. The behind smokebox configuration has the advantage that a good connection can be made to the steam pipe and was used by the Great Western Railway. It has the disadvantage that the lubricator's accessibility is reduced and additional drain pipes are required to be connected to the waste to avoid it dripping onto the boiler.

Later types of lubricator referred to as the "sight-feed" type, allowed a sight glass to be positioned in the cab where the rate of oil feed could be observed.

The displacement lubricator was a useful stop-gap but had the disadvantage that it was difficult to accurately control over the rate of oil feed and lubrication was only supplied when the engine was doing work (when a locomotive is coasting with the regulator closed, no steam is present to operate the lubricator). In the Ramsbottom type the flow of lubricant depended mainly on the weather. In cold weather the rate of feed would increase because steam would condense more rapidly.

Roscoe Type

The Roscoe type lubricator improved the situation by providing a valve with which to regulate the flow of steam and hence lubricant. Setting the valve correctly required experience and depended on the speed of the train.

The Roscoe lubricator was invented by James Roscoe in 1862 and patented in British Patent 1337. It incorporated two improvements over the original Ramsbottom type, the ability to control the quantity of steam which entered the lubricator, via a control valve, and the addition of an air filled chamber within the oil reservoir. The chamber expanded when steam was cut off, the aim of which was to supply oil even when the locomotive was coasting. However, this proved ineffective in practice, particularly if the locomotive was coasting for a long time (e.g. down a hill).

Replacement

When more sophisticated lubricators, such as the Wakefield and Detroit types, were developed, displacement lubricators fell out of use but they are still used on model steam engines.

Detroit Hydrostatic Lubricator

This type of hydrostatic displacement lubricator was made by the Detroit Lubricator Company of the USA. A British patent was granted in 1911 and this lubricator was then manufactured by the Vacuum Oil Company (later Mobil Oil) as the British Detroit Lubricator. This lubricator was commonly made with either 3,4 or 5 feeds, the center feed on the odd numbered lubricators was often used for supplying oil to the steam end of Westinghouse Air compressors.

The one piece lubricator body is a gunmetal casting and incorporates the condensing chamber, the oil reservoir and the sight glasses. Passages within the body interconnect these areas. The body is fitted with a steam valve, a water valve, an oil control valve. Sight glasses are fitted at the front & back of the body on each of the feeds, these feeds are fitted with individual feed regulating valves for controlling the feed rate for each outlet. The oil control valve was used to either turn the supply off/on, or to shut off the supply to the cylinders and just supply the air compressor with oil whilst the locomotive is stationary on lubricators with a center feed. Each feed is also fitted with a drain valve for cleaning each sight glass. The body is also fitted with an oil chamber drain for draining the condensed water from the oil chamber.

Silvertown Mechanical Lubricator

The Silvertown lubricator was designed and developed by the Midland Railway in 1911 and from 1922 was commercially available from Gresham and Craven. This style of mechanical lubricator was used by the London, Midland and Scottish Railway and latter by British Rail, as well as by the various UK private locomotive builders for export overseas. Commonly 2 lubricators were fitted, one for lubricating the cylinders and the second for lubricating the axleboxes.

The lubricator is a self-contained unit with the drive mechanism fitted at one end, with the drive shaft running through the cast iron oil reservoir. The drive shaft is fitted with

2 hardened steel eccentrics which provide movement to the drive frame which runs in machined grooves in the oil reservoir. This drive frame reciprocates vertically which then drives the double acting independent oil pumps located in the oil reservoir.

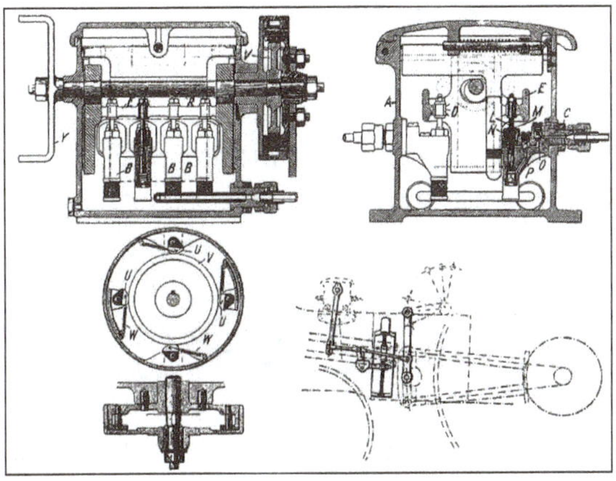

Patent drawing of the original Midland Railway version of the Silverton mechanical lubricator.

The lubricator is driven by a rod from the expansion link in locomotives valve gear. The rocking motion of this drive is converted to a rotary motion by a series of driving pawls contained in a clutch box. The lubricators were also fitted with a priming hand wheel on the outer face of the lubricator.

Friedmann/Nathan Mechanical Lubricator

The Friedmann system of mechanical lubricators included the DV & FSA type lubricators. This style of mechanical lubricator was designed & produced by Alex Friedmann KG of Austria.

Both types of lubricator are of the 'valveless' type, having no ball valves, springs or oil seals used in the operation of the pump units.

The DV range was available in range of sizes with capacities ranging from 8 to 36 pints, and the number of feeds from 1 to 26. Each of the lubricators consist of a reservoir, a ratchet assembly, a sliding shaft assembly, and the pump units. A screwed oil filler cap is fitted to the lockable hinged lid of the reservoir and a sight glass for indicating the oil level is fitted to the front face of the reservoir. The pumps of the DV type lubricator are operated by a sliding shaft assembly, unlike the Silverton or Wakefield lubricators which are operated by a rotating camshaft.

The FSA type lubricator differs by each pump having two separate pistons to control the pumping & distribution of oil. The pistons are driven by cams on the driving shaft. The driving shaft is driven via a means of a roller ratchet assembly which converts the reciprocating motion of the operating linkage to rotary motion.

Basic Modes of Lubrication

In lubrication, two contacting surfaces which have motion relative to each other are seperated by a high pressure film of lubricant.

Based on the method of creating high pressure film of lubricant between the contacting surfaces, the lubrication can be classified into following four groups:

- Hydrodynamic lubrication.
- Hydrostatic lubrication.
- Elastohydrodynamic lubrication.
- Solid film lubrication.

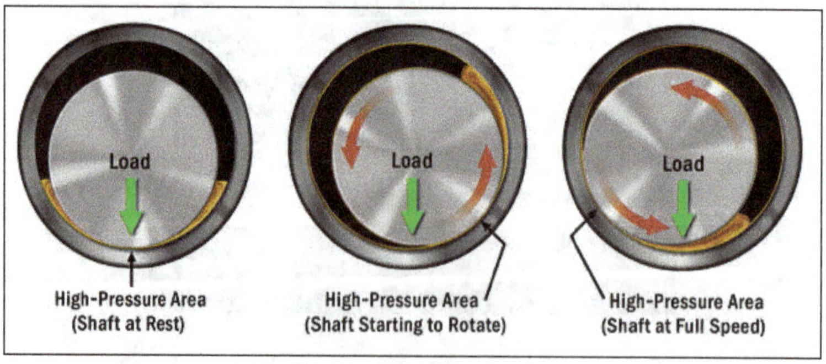

Hydrodynamic Lubrication.

As shown in figure, in hydrodynamic lubrication, the two surfaces in relative motion are separated by a relatively thick fluid film to prevent the metal to metal contact. In this type of lubrication, the load supporting high pressure fluid film is created due to mainly two reasons:

- The shape of the zone between the contacting surfaces.
- Relative motion between the contacting surfaces.

The moving surface pulls the lubricant into a wedge shaped zone at a velocity sufficiently high to create the high pressure film, necessary to separate the two surfaces against the load. The hydrodynamic lubrication is also known as thick film lubrication or full film lubrication.

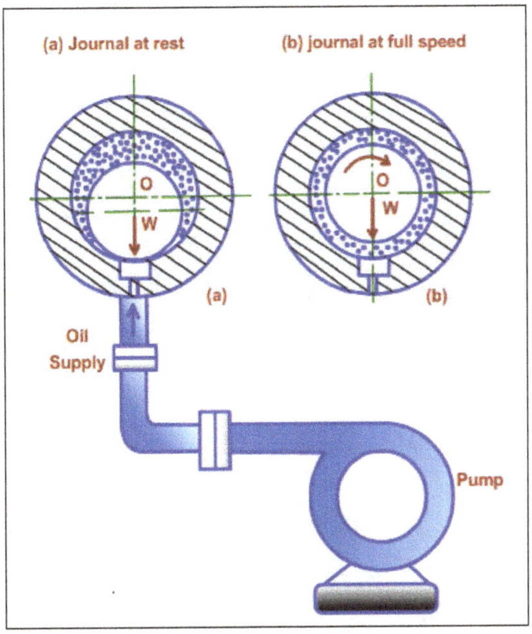

Hydrostatic Lubrication.

In hydrostatic lubrication, the load supporting high pressure fluid film is created by an external source like pump. The lubrication which is pressurized externally is supplied between the two surfaces. So, in hydrostatic lubrication the motion of one surface relative to other is not required.

This type of lubrication occurs when a lubricant is introduced between the surfaces which in rolling contact, such as mating of gears or rolling elements and races of the rolling contact bearing. The lubricant is entrapped in the deformation zone between the contacting surfaces.

When the bearings are required to operate at very high temperatures, the lubricating oils are not suitable as they may lose their properties at such high temperatures. In such cases, solid film lubricant, such as molybdenum disulphide or graphite are used.

Boundary Lubrication

Boundary lubrication is lubrication by a liquid under conditions where the solid surfaces are so close together that appreciable contact between opposing asperities is possible. The friction and wear in boundary lubrication are determined predominantly by interaction between the solids and between the solids and the liquid. The bulk flow properties of the liquid play little or no part in the friction and wear behavior.

Based on friction experiments on bearings, Stribeck expressed the relationship between the frictioncoefficient [f], viscosity of the lubricating oil [η], load [F_N], and velocity [V] in the Stribeck curve. This curve brilliantly captures the characteristics of various lubrication regions, including [I] boundary lubrication, [II] elastohydrodynamic lubrication (EHL), and mixed lubrication.

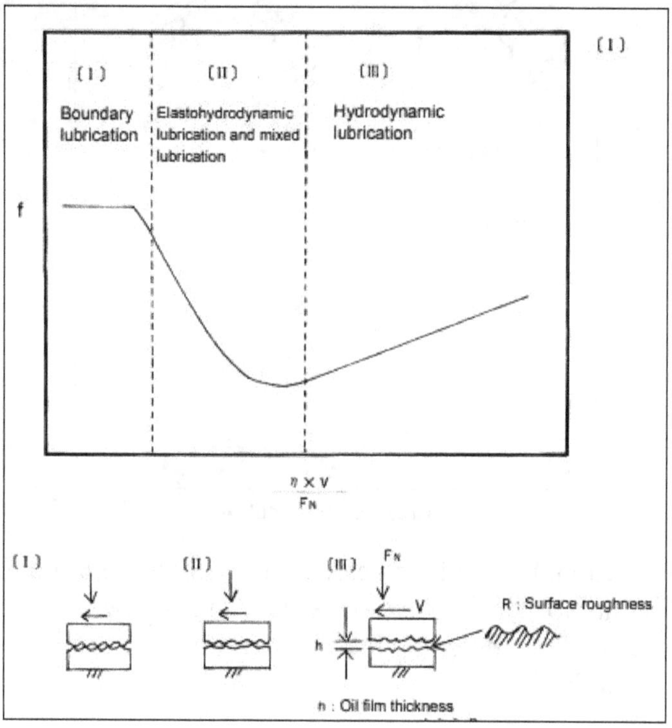

Stribeck curve and lubrication regions.

In hydrodynamic lubrication, the fluid completely isolates the friction surfaces [h >> R], and internal fluid friction alone determines tribological characteristics. In elastohydrodynamic lubrication [h ≈ R], fluid viscosity, the viscosity-pressure coefficient and the elastic coefficient of the solid surface are the most dominant factors. In contrast, the boundary lubrication mode is mainly characterized by the following three points:

- Friction surfaces are in contact at microasperities.
- Hydrodynamic effects of lubricating oil or rheological characteristics of bulk do not significantly influence tribological characteristics.
- Interactions in the contact between friction surfaces and between friction surfaces and the lubricant (including additives) dominate tribological characteristics.

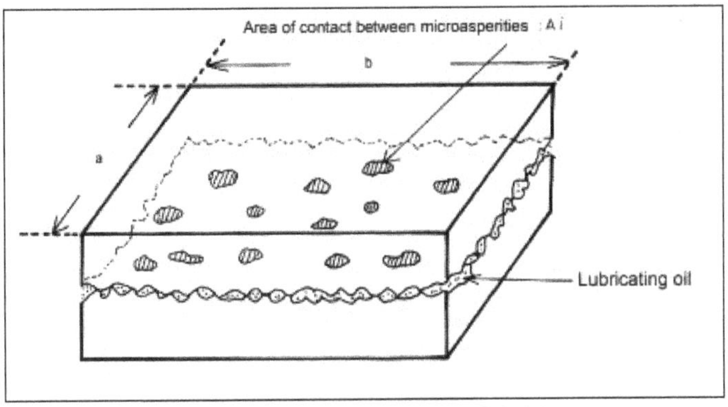

Friction surface contact model (A).

In other words, since the thickness [h] of the oil film between the friction surfaces is smaller than their surface roughness [R] in boundary lubrication, metallic contact occurs across every portion of the surface, as shown in figure. Friction surfaces contact each other over a true contact area ($\sum_{i=1}^{n} Ai$, n: number of contact points), which is the sum of the contact areas [Ai] between the discontinuous microasperities, and the load is supported by a much smaller area than the apparent contact area (a × b). Thus, in boundary lubrication where large amounts of friction increase the temperatures of friction surfaces and cause wear and seizure and various resulting problems, it is critical to reduce friction and wear by preventing metallic contact; that is, it is critical to form lubricating films on friction surfaces by applying lubricants (lubricating oils, additives, and solid lubricants).

As shown in figure (the cross section of figure), when oil film rupture causes metallic contact, friction force [F] is spent on the shearing not only of the oil film, but of metal bonding. Thus, F is expressed by the following equation:

$$F = A\{\alpha Sm + (1-\alpha)S_i\}$$

Where A is load bearing area, α is the ratio of metallic contact, S_m is the shear strength of metal bonding, and S_l is the shear strength of the oil film. Since S_l is significantly smaller than S_m, minimizing α reduces friction.

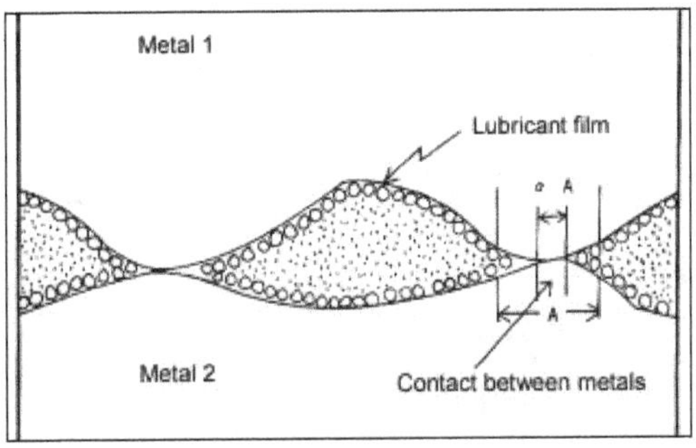

Friction surface contact model (B).

Wear in Boundary Lubrication

The main types of wear in boundary lubrication are as follows:

- Adhesive wear: Caused by adhesion of microasperities to friction surfaces.
- Abrasive wear: Caused by cutting by hard, foreign particles or abrasive particles in oil.
- Surface fatigue wear: Caused by surface fatigue resulting from repeated stress and tribological chemical interactions.
- Chemical wear: Dominated by chemical reactions between friction surfaces and lubricant molecules.

In most cases, these types of wear more or less occur side by side. Mechanical conditions, types of lubricants and metal materials, oil temperatures, atmosphere, generation of friction heat, exo-electron emission, and other factors result in a wide range of complex phenomena on friction surfaces, making boundary lubrication problematic.

As shown in figure, Solid and solid are in relative motion under given mechanical conditions (load and sliding speed). In the area of contact between the two solids, adhesion and mutual transfer cause adhesive wear, while generated abrasive particles or hard foreign particles cause abrasive wear. In contrast, interactions between solid and lubricant or atmosphere (for example, adsorption and chemical reaction) form a lubricating film that reduces friction and wear and prevents seizure. However, chemical reaction products are sometimes removed into the oil by shearing and dissolution, causing chemical wear.

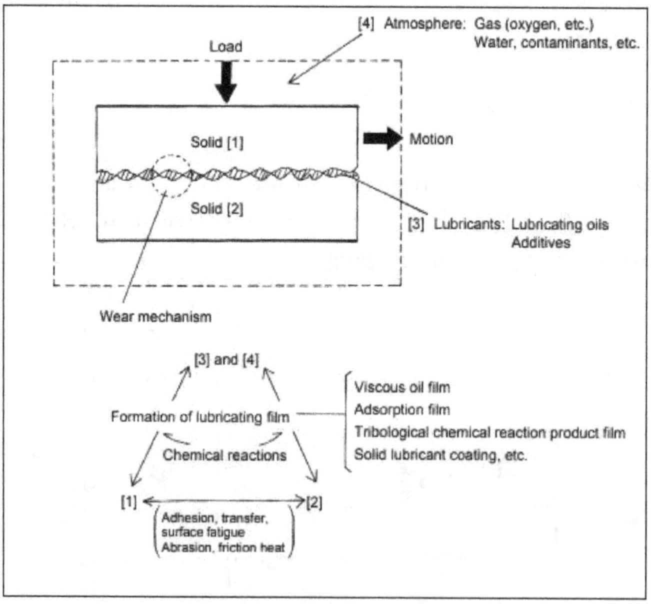

Wear under boundary lubrication.

Lubrication Theory

In fluid dynamics, lubrication theory describes the flow of fluids (liquids or gases) in a geometry in which one dimension is significantly smaller than the others. An example is the flow above air hockey tables, where the thickness of the air layer beneath the puck is much smaller than the dimensions of the puck itself.

Internal flows are those where the fluid is fully bounded. Internal flow lubrication theory has many industrial applications because of its role in the design of fluid bearings. Here a key goal of lubrication theory is to determine the pressure distribution in the fluid volume, and hence the forces on the bearing components. The working fluid in this case is often termed a lubricant.

Free film lubrication theory is concerned with the case in which one of the surfaces containing the fluid is a free surface. In that case the position of the free surface is itself unknown, and one goal of lubrication theory is then to determine this. Surface tension may then be significant, or even dominant. Issues of wetting and dewetting then arise. For very thin films (thickness less than one micrometre), additional intermolecular forces, such as Van der Waals forces or disjoining forces, may become significant.

A thin layer of liquid mixed with particles flowing down an inclined plane.

Theoretical Basis

Mathematically, lubrication theory can be seen as exploiting the disparity between two length scales. The first is the characteristic film thickness, H, and the second is a characteristic substrate length scale L. The key requirement for lubrication theory is that the ratio $\varepsilon = H / L$ is small, that is, $\epsilon \ll 1$. The Navier–Stokes equations (or Stokes equations, when fluid inertia may be neglected) are expanded in this small parameter, and the leading-order equations are then,

$$\frac{\partial p}{\partial x}=0$$

$$\frac{\partial p}{\partial z}=\mu\frac{\partial^2 u}{\partial z^2}$$

Where x and z are coordinates in the direction of the substrate and perpendicular to it respectively. Here p is the fluid pressure, and u is the fluid velocity component parallel to the substrate; μ is the fluid viscosity. The equations show, for example, that pressure variations across the gap are small, and that those along the gap are proportional to the fluid viscosity. A more general formulation of the lubrication approximation would include a third dimension, and the resulting differential equation is known as the Reynolds equation.

Applications

An important application area is lubrication of machinery components such as fluid bearings and mechanical seals. Coating is another major application area including the preparation of thin films, printing, painting and adhesives.

Biological applications have included studies of red blood cells in narrow capillaries and of liquid flow in the lung and eye.

Theory of Hydrodynamic Lubrication

Hydrodynamic lubrication is a way that is used to reduce friction and/or wear of rubbing solids with the aid of liquid (or semi-solid) lubricant. For a vast majority of the surfaces encountered in nature and used in industry, the source of friction is the imperfections of the surfaces. Even mirror shining surfaces are composed of hills and valleys – surface roughness. The goal of hydrodynamic lubrication is to add a proper lubricant, so that it penetrates into the contact zone between rubbing solids and creates a thin liquid film, This film separates the surfaces from direct contact and it in general reduces friction and consequently wear (but not always), since friction within the lubricant is less than between the directly contacting solids. Hydrodynamic theory is developed within a field known as tribology.

Lubricant is a substance which is used to control (more often to reduce) friction and wear of the surfaces in a contact of the bodies in relative motion. Depending on its nature, lubricants are also used to eliminate heat and wear debris, supply additives into the contact, transmit power, protect, seal. A lubricant can be in liquid (oil, water, etc.), solid (graphite, graphene, molybdenum disulfide), gaseous (air) or even semisolid (grease) forms. Most of the lubricants contain additives (5-30%) to improve their performance.

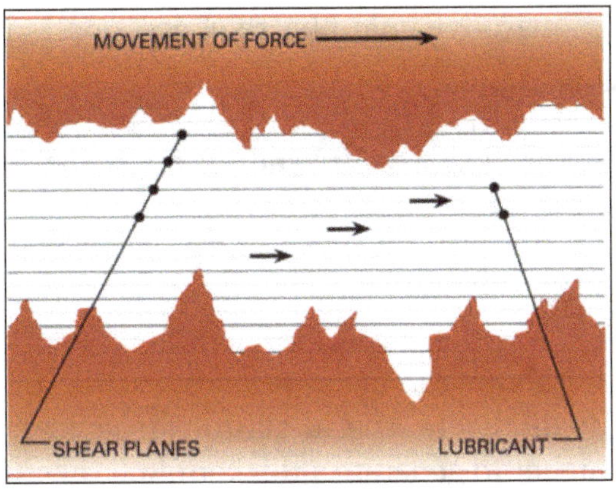

Lubricant Film.

History of lubrication theory goes more than a century back to 1886 when O. Reynolds published famous equation of thin fluid film flow in the narrow gap between two solids. This equation carries his name and forms a foundation of the lubrication theory.

It differs from elastohydrodynamic lubrication theory (EHL) due to negligible elastic deformation of the surfaces. Absence of the elastic deformations simplifies the problem as compared to EHL theory, but allows one to construct some important analytical solutions.

Solutions of Reynolds Equation

First solutions of lubrication theory were obtained by Reynolds himself and can be found in the original work. One of the most important analytical solutions of the hydrodynamic theory (of the most interest in tribology field) was obtained by Martin in 1916. The solution considers a relative motion of cylinder on flat plane.

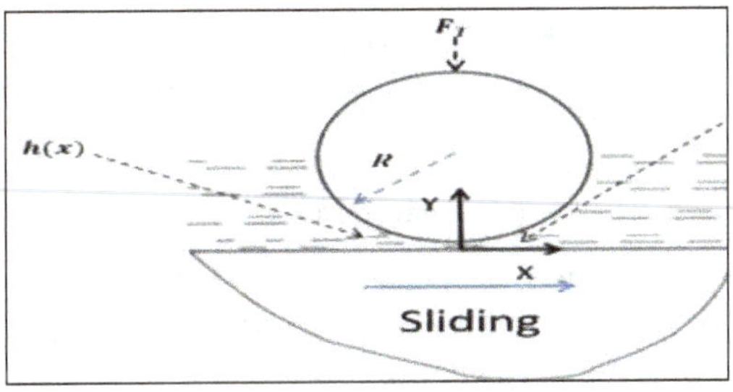

Following system of equations is considered:

$$\frac{\partial}{\partial x}\left(\frac{h^3}{\mu}\frac{\partial p}{\partial x}\right) = 6U_m \frac{\partial h}{\partial x}$$

$$h(x) = h_0 + \frac{x^2}{2R}$$

$$F_l = \int_{-\infty}^{\infty} p\,dx$$

Where h, p, μ, α, U_m are hydrodynamic film thickness, pressure, viscosity and average sliding speed correspondingly ($U_m = \frac{U_1 + U_2}{2}$). F_l here is the normal load (per unit length). There are two unknowns, pressure and the approach h_o with two equations to determine them. Martin's solution states following:

$$h_o = 4.9 \frac{\mu U_m R}{F_l}$$

This solution immediately shows the major relations within the hydrodynamic theory (but they also remain qualitatively true in elastohydrodynamic theory as well). The sliding speed has to be higher to generate a thicker film. The same is true for viscosity: higher viscosity leads to a thicker lubricant film. It should be noted, that it is typically desired to have a sufficiently thick lubricant film, so that the surface are completely separated to reduce wear. At the same time, it is clear, that the friction (hydrodynamic) will increase with the increase of both, sliding speed and viscosity. This in turn leads to energy losses. Therefore, there will be a trade-off between the wear performance and the optimization of energy losses.

Currently, continuous efforts are undertaken to reduce the energy losses and to move towards a sustainable society and at the same time increase the reliability of tribological devices. According to the discussion there is a contradiction between the wear performance and the energetic performance. Therefore, classical lubrication theory has reached its fundamental limit in the energy losses reduction and new theories have to be developed. From that standpoint, solid lubricants, such as graphene or diamond like carbon are promising materials to reduce friction further and reach a so-called superlubricity regime.

Reynolds Equations and its Limitations

The Reynolds Equation is a partial differential equation governing the pressure distribution of thin viscous fluid films in Lubrication theory. It should not be confused with Osborne Reynolds' other namesakes, Reynolds number and Reynolds-averaged Navier–Stokes equations. It was first derived by Osborne Reynolds in 1886. The classical Reynolds Equation can be used to describe the pressure distribution in nearly any type of fluid film bearing; a bearing type in which the bounding bodies are fully separated by a thin layer of liquid or gas.

General Usage

The general Reynolds equation is:

$$\frac{\partial}{\partial x}\left(\frac{\rho h^3}{12\mu}\frac{\partial p}{\partial x}\right) + \frac{\partial}{\partial y}\left(\frac{\rho h^3}{12\mu}\frac{\partial p}{\partial y}\right)$$

$$= \frac{\partial}{\partial x}\left(\frac{\rho h(u_a + u_b)}{2}\right) + \frac{\partial}{\partial y}\left(\frac{\rho h(v_a + v_b)}{2}\right) + \rho(w_a - w_b) - \rho u_a \frac{\partial h}{\partial x} - \rho v_a \frac{\partial h}{\partial y} + h\frac{\partial \rho}{\partial t}$$

Where:

- p is fluid film pressure.
- X and y are the bearing width and length coordinates.
- z is fluid film thickness coordinate.
- h is fluid film thickness.
- μ is fluid viscosity.
- ρ is fluid density.
- u, v, w are the bounding body velocities in x, y, z respectively.
- a, b are subscripts denoting the top and bottom bounding bodies respectively.

The equation can either be used with consistent units or nondimensionalized.

The Reynolds Equation assumes:

- The fluid is Newtonian.
- Fluid viscous forces dominate over fluid inertia forces. This is the principal of the Reynolds number.
- Fluid body forces are negligible.
- The variation of pressure across the fluid film is negligibly small $\left(\text{i.e.} \frac{\partial p}{\partial z} = 0\right)$.
- The fluid film thickness is much less than the width and length and thus curvature effects are negligible. (i.e. $h \ll l$ and $h \ll w$).

For some simple bearing geometries and boundary conditions, the Reynolds equation can be solved analytically. Often however, the equation must be solved numerically. Frequently this involves discretizing the geometric domain, and then applying a finite technique - often FDM, FVM, or FEM.

Solution of Reynolds Equation

In general, Reynolds equation has to be solved using numerical methods such as finite difference, or finite element. In certain simplified cases, however, analytical or approximate solutions can be obtained.

For the case of rigid sphere on flat geometry, steady-state case and half-Sommerfeld cavitation boundary condition, the 2-D Reynolds equation can be solved analytically. This solution was proposed by a Nobel Prize winner Pyotr Kapitsa. Half-Sommerfeld boundary condition was shown to be inaccurate and this solution has to be used with care.

In case of 1-D Reynolds equation several analytical or semi-analytical solutions are available. In 1916 Martin obtained a closed form solution for a minimum film thickness and pressure for a rigid cylinder and plane geometry. This solution is not accurate for the cases when the elastic deformation of the surfaces contributes considerably to the film thickness. In 1949, Grubin obtained an approximate solution for so called elasto-hydrodynamic lubrication (EHL) line contact problem, where he combined both elastic deformation and lubricant hydrodynamic flow. In this solution it was assumed that the pressure profile follows Hertz solution. The model is therefore accurate at high loads, when the hydrodynamic pressure tends to be close to the Hertz contact pressure.

Applications

The Reynolds equation is used to model the pressure in many applications. For example:

- Ball bearings,
- Air bearings,
- Journal bearings,
- Squeeze film dampers in aircraft gas turbines,
- Human hip and knee joints,
- Lubricated gear contacts.

Reynolds Equation Adaptations

In 1978 Patir and Cheng introduced an average flow model which modifies the Reynolds equation to consider the effects of surface roughness on partially lubricated contacts.

Viscosity Index

The viscosity index (VI) is an arbitrary, unitless measure of the change of viscosity with temperature, mostly used to characterize the viscosity-temperature behavior of lubricating oils. The lower the VI, the more the viscosity is affected by changes in temperature.

The VI was originally measured on a scale from 0 to 100; however, advancements in lubrication science have led to the development of oils with much higher VIs.

The viscosity of a lubricant is closely related to its ability to reduce friction in solid body contacts. Generally, the least viscous lubricant which still forces the two moving surfaces apart to achieve "fluid bearing" conditions is desired. If the lubricant is too viscous, it will require a large amount of energy to move (as in honey); if it is too thin, the surfaces will come in contact and friction will increase.

Many lubricant applications require the lubricant to perform across a wide range of conditions, for example, automotive lubricants are required to reduce friction between engine components when the engine is started from cold (relative to the engine's operating temperatures) up to 200 °C or 392 °F when it is running. The best oils with the highest VI will remain stable and not vary much in viscosity over the temperature range. This allows for consistent engine performance within the normal working conditions. Historically, there were two different oil types recommended for usage in different weather conditions. As an example, with winter oils and cold starting the engines, and with temperature ranges from, say, -30 °C to 0 °C, a 5 weight oil would be pumpable at the very low temperatures and the generally cooler engine operating temperatures. However, in hot climates where temperatures range from 30 °C to 45 °C, a 50 weight oil would be necessary so it would remain thick enough to hold up an oil film between the moving hot parts.

Thus the issue of multigrade oils came into being, where with variable temperatures of, say, minus -10 °C during the cold nights and 20 °C during the days, a 5 weight oil would be good as the oil would be pumpable in a cold engine and as the engine came up to running temperature, and the day warmed up, the characteristics of a 30 weight oil would be ideal. Thus the 5W-30W or 5-30W oils were introduced, rather than the fixed and temperature limiting grades — where the thin oils became too thin when hot and the thicker oils became too thick when cold.

This effects of temperature on a single viscosity oil can be demonstrated by pouring a small amount of vegetable oil into a pot or pan and then either cooling it in a freezer or heating it on a cooking stove. When oils get cold enough in a deep freezer, they will solidify into a block of "wax" like oil that cannot be pumped around inside an engine's lubrication system. However when a spoonful of very cold oil is put into a pan on a stove and it is slowly heated and swirled around, the oil will gradually warm up, and there is a definite temperature range where the oil is warm and very "oily" or quite lubricating. However as the oil is heated further, the oil becomes thinner and thinner until it is nearly smoking and it is almost as thin as water — and thus it has almost no capacity to keep moving parts separated — resulting in metal to metal contact and damage of the components that are supposed to be kept apart with a thin film of oil.

Thus the multigrade oils are recommended for use based on the ambient temperature ranges of the season or environment.

Additionally, there are the issues of oil temperature maintenance, such as oil or engine heaters that enable easy starting and shorter warm up period in very cold climates, and oil coolers to dump enough heat from the oil, and thus the engine, gearbox, or hydraulic oil circuit, so as to keep the oil's upper temperature to within a specified upper working limit.

It is wise to follow the manufacturers recommendation as to what grades of oil or types of lubricants should be used in or on any equipment relative to the seasonal temperature ranges of the environment that the equipment is used in.

Despite all the neurosis generating literature of the advertising industry, a fairly simple oil viscosity and temperature chart does fulfill the educational needs of the user and the enquipment, the latter being reasonably typical of all owner manuals and combustion engine manufacturers. As stipulated earlier, always follow the specifications of your equipment manufacturer.

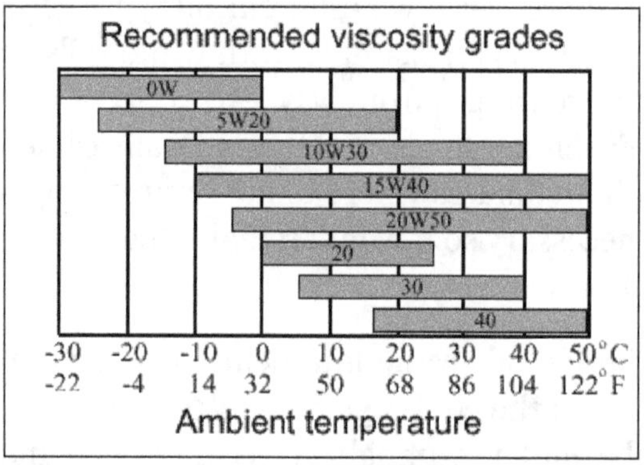

Typical Engine Oil Temperature Viscosity Chart.

Classification

The VI scale was set up by the Society of Automotive Engineers (SAE). The temperatures chosen arbitrarily for reference are 100 and 210 °F (38 and 99 °C). The scale was originally interpolated between 0 for a naphthenic Texas Gulf crude and 100 for a paraffinnic Pennsylvania crude. Since the inception of the scale better oils have also been produced, leading to VIs greater than 100.

VI improving additives and higher quality base oils are widely used nowadays which increase the VIs attainable beyond the value of 100. The Viscosity Index of synthetic oils ranges from 80 to over 400.

Viscosity index	Classification
Under 35	Low
35 to 80	Medium
80 to 110	High
Above 110	Very high

Calculation

The viscosity index can be calculated using the following formula:

$$VI = 100\frac{L-U}{L-H}$$

where U is the oil's kinematic viscosity at 40 °C (104 °F), and L and H are values based on the oil's kinematic viscosity at 100 °C (212 °F). L and H are the values of viscosity at 40 °C for oils of VI 0 and 100 respectively, having the same viscosity at 100 °C as the oil whose VI we are trying to determine. These L and H values can be found in ASTM D2270.

Lubricants

A lubricant is a substance, usually organic, introduced to reduce friction between surfaces in mutual contact, which ultimately reduces the heat generated when the surfaces move. It may also have the function of transmitting forces, transporting foreign particles, or heating or cooling the surfaces. The property of reducing friction is known as lubricity.

In addition to industrial applications, lubricants are used for many other purposes. Other uses include cooking (oils and fats in use in frying pans, in baking to prevent food sticking), bioapplications on humans (e.g. lubricants for artificial joints), ultrasound examination, medical examination. It is mainly used to reduce friction and to contribute to a better and efficient functioning of a mechanism.

Properties

A good lubricant generally possesses the following characteristics:

- A high boiling point and low freezing point (in order to stay liquid within a wide range of temperature).
- A high viscosity index.
- Thermal stability.
- Hydraulic stability.
- Demulsibility.
- Corrosion prevention.
- A high resistance to oxidation.

Formulation

Typically lubricants contain 90% base oil (most often petroleum fractions, called mineral oils) and less than 10% additives. Vegetable oils or synthetic liquids such as hydrogenated polyolefins, esters, silicones, fluorocarbons and many others are sometimes used as base oils. Additives deliver reduced friction and wear, increased viscosity, improved viscosity index, resistance to corrosion and oxidation, aging or contamination, etc.

Non-liquid lubricants include powders (dry graphite, PTFE, molybdenum disulphide, tungsten disulphide, etc), PTFE tape used in plumbing, air cushion and others. Dry lubricants such as graphite, molybdenum disulphide and tungsten disulphide also offer lubrication at temperatures (up to 350 °C) higher than liquid and oil-based lubricants are able to operate. Limited interest has been shown in low friction properties of compacted oxide glaze layers formed at several hundred degrees Celsius in metallic sliding systems, however, practical use is still many years away due to their physically unstable nature.

Additives

A large number of additives are used to impart performance characteristics to the lubricants. Modern automotive lubricants contain as many as ten additives, comprising up to 20% of the lubricant, the main families of additives are:

- Pour point depressants are compounds that prevent crystallization of waxes. Long chain alkylbenzenes adhere to small crystallites of wax, preventing crystal growth.
- Anti-foaming agents are typically silicone compounds which increase surface tension in order to discourage foam formation.
- Viscosity index improvers (VIIs) are compounds that allow lubricants to remain viscous at higher temperatures. Typical VIIs are polyacrylates and butadiene.
- Antioxidants suppress the rate of oxidative degradation of the hydrocarbon molecules within the lubricant. At low temperatures, free radical inhibitors such as hindered phenols are used, e.g. butylated hydroxytoluene. At temperatures >90 °C, where the metals catalyze the oxidation process, dithiophosphates are more useful. In the latter application the additives are called metal deactivators.
- Detergents ensure the cleanliness of engine components by preventing the formation of deposits on contact surfaces at high temperatures.
- Corrosion inhibitors (rust inhibitors) are usually alkaline materials, such as alkylsulfonate salts, that absorb acids that would corrode metal parts.
- Anti-wear additives form protective 'tribofilms' on metal parts, suppressing

wear. They come in two classes depending on the strength with which they bind to the surface. Popular examples include phosphate esters and zinc dithiophosphates.

- Extreme pressure (anti-scuffing) additives form protective films on sliding metal parts. These agents are often sulfur compounds, such as dithiophosphates.
- Friction modifiers reduce friction and wear, particularly in the boundary lubrication regime where surfaces come into direct contact.

Functions of Lubricants

One of the largest applications for lubricants, in the form of motor oil, is protecting the internal combustion engines in motor vehicles and powered equipment.

Lubricant vs. Anti-tack Coating

Anti-tack or anti-stick coatings are designed to reduce the adhesive condition (stickiness) of a given material. The rubber, hose, and wire and cable industries are the largest consumers of anti-tack products but virtually every industry uses some form of anti-sticking agent. Anti-sticking agents differ from lubricants in that they are designed to reduce the inherently adhesive qualities of a given compound while lubricants are designed to reduce friction between any two surfaces.

Keep Moving Parts Apart

Lubricants are typically used to separate moving parts in a system. This separation has the benefit of reducing friction, wear and surface fatigue, together with reduced heat generation, operating noise and vibrations. Lubricants achieve this in several ways. The most common is by forming a physical barrier i.e., a thin layer of lubricant separates the moving parts. This is analogous to hydroplaning the loss of friction observed when a car tire is separated from the road surface by moving through standing water. This is termed hydrodynamic lubrication. In cases of high surface pressures or temperatures, the fluid film is much thinner and some of the forces are transmitted between the surfaces through the lubricant.

Reduce Friction

Typically the lubricant-to-surface friction is much less than surface-to-surface friction in a system without any lubrication. Thus use of a lubricant reduces the overall system friction. Reduced friction has the benefit of reducing heat generation and reduced formation of wear particles as well as improved efficiency. Lubricants may contain additives known as friction modifiers that chemically bind to metal surfaces to reduce surface friction even when there is insufficient bulk lubricant present for hydrodynamic lubrication, e.g. protecting the valve train in a car engine at startup.

Transfer Heat

Both gas and liquid lubricants can transfer heat. However, liquid lubricants are much more effective on account of their high specific heat capacity. Typically the liquid lubricant is constantly circulated to and from a cooler part of the system, although lubricants may be used to warm as well as to cool when a regulated temperature is required. This circulating flow also determines the amount of heat that is carried away in any given unit of time. High flow systems can carry away a lot of heat and have the additional benefit of reducing the thermal stress on the lubricant. Thus lower cost liquid lubricants may be used. The primary drawback is that high flows typically require larger sumps and bigger cooling units. A secondary drawback is that a high flow system that relies on the flow rate to protect the lubricant from thermal stress is susceptible to catastrophic failure during sudden system shut downs. An automotive oil-cooled turbocharger is a typical example. Turbochargers get red hot during operation and the oil that is cooling them only survives as its residence time in the system is very short (i.e. high flow rate). If the system is shut down suddenly (pulling into a service area after a high-speed drive and stopping the engine) the oil that is in the turbo charger immediately oxidizes and will clog the oil ways with deposits. Over time these deposits can completely block the oil ways, reducing the cooling with the result that the turbo charger experiences total failure, typically with seized bearings. Non-flowing lubricants such as greases and pastes are not effective at heat transfer although they do contribute by reducing the generation of heat in the first place.

Carry Away Contaminants and Debris

Lubricant circulation systems have the benefit of carrying away internally generated debris and external contaminants that get introduced into the system to a filter where they can be removed. Lubricants for machines that regularly generate debris or contaminants such as automotive engines typically contain detergent and dispersant additives to assist in debris and contaminant transport to the filter and removal. Over time the filter will get clogged and require cleaning or replacement, hence the recommendation to change a car's oil filter at the same time as changing the oil. In closed systems such as gear boxes the filter may be supplemented by a magnet to attract any iron fines that get created.

It is apparent that in a circulatory system the oil will only be as clean as the filter can make it, thus it is unfortunate that there are no industry standards by which consumers can readily assess the filtering ability of various automotive filters. Poor automotive filters significantly reduce the life of the machine (engine) as well as making the system inefficient.

Transmit Power

Lubricants known as hydraulic fluid are used as the working fluid in hydrostatic power transmission. Hydraulic fluids comprise a large portion of all lubricants produced in the world. The automatic transmission's torque converter is another important application for power transmission with lubricants.

Protect against Wear

Lubricants prevent wear by keeping the moving parts apart. Lubricants may also contain anti-wear or extreme pressure additives to boost their performance against wear and fatigue.

Prevent Corrosion

Many lubricants are formulated with additives that form chemical bonds with surfaces or that exclude moisture, to prevent corrosion and rust. It reduces corrosion between two metallic surface and avoids contact between these surfaces to avoid immersed corrosion.

Seal for Gases

Lubricants will occupy the clearance between moving parts through the capillary force, thus sealing the clearance. This effect can be used to seal pistons and shafts.

Fluid Types

- Automotive:
    - Engine oils:
        - Petrol (Gasolines) engine oils.
        - Diesel engine oils.
    - Automatic transmission fluid.
    - Gearbox fluids.
    - Brake fluids.
    - Hydraulic fluids.
- Tractor (one lubricant for all systems):
    - Universal Tractor Transmission Oil – UTTO.
    - Super Tractor Oil Universal – STOU – includes engine.
- Other motors:
    - 2-stroke engine oils.

- Industrial:
  - Hydraulic oils.
  - Air compressor oils.
  - Food Grade lubricants.
  - Gas Compressor oils.
  - Gear oils.
  - Bearing and circulating system oils.
  - Refrigerator compressor oils.
  - Steam and gas turbine oils.
- Aviation:
  - Gas turbine engine oils.
  - Piston engine oils.
- Marine:
  - Crosshead cylinder oils.
  - Crosshead Crankcase oils.
  - Trunk piston engine oils.
  - Stern tube lubricants.

Glaze Formation (High Temperature Wear)

A further phenomenon that has undergone investigation in relation to high temperature wear prevention and lubrication, is that of a compacted oxide layer glaze formation. Such glazes are generated by sintering a compacted oxide layer. Such glazes are crystalline, in contrast to the amorphous glazes seen in pottery. The required high temperatures arise from metallic surfaces sliding against each other (or a metallic surface against a ceramic surface). Due to the elimination of metallic contact and adhesion by the generation of oxide, friction and wear is reduced. Effectively, such a surface is self-lubricating.

As the "glaze" is already an oxide, it can survive to very high temperatures in air or oxidising environments. However, it is disadvantaged by it being necessary for the base metal (or ceramic) having to undergo some wear first to generate sufficient oxide debris.

Disposal and Environmental Impact

It is estimated that 40% of all lubricants are released into the environment. Common disposal methods include recycling, burning, landfill and discharge into water, though typically disposal in landfill and discharge into water are strictly regulated in most countries, as even small amount of lubricant can contaminate a large amount of water. Most regulations permit a threshold level of lubricant that may be present in waste streams and companies spend hundreds of millions of dollars annually in treating their waste waters to get to acceptable levels.

Burning the lubricant as fuel, typically to generate electricity is also governed by regulations mainly on account of the relatively high level of additives present. Burning generates both airborne pollutants and ash rich in toxic materials, mainly heavy metal compounds. Thus lubricant burning takes place in specialized facilities that have incorporated special scrubbers to remove airborne pollutants and have access to landfill sites with permits to handle the toxic ash.

Unfortunately, most lubricant that ends up directly in the environment is due to general public discharging it onto the ground, into drains and directly into landfills as trash. Other direct contamination sources include runoff from roadways, accidental spillages, natural or man-made disasters and pipeline leakages.

Improvement in filtration technologies and processes has now made recycling a viable option (with rising price of base stock and crude oil). Typically various filtration systems remove particulates, additives and oxidation products and recover the base oil. The oil may get refined during the process. This base oil is then treated much the same as virgin base oil however there is considerable reluctance to use recycled oils as they are generally considered inferior. Basestock fractionally vacuum distilled from used lubricants has superior properties to all natural oils, but cost effectiveness depends on many factors. Used lubricant may also be used as refinery feedstock to become part of crude oil. Again, there is considerable reluctance to this use as the additives, soot and wear metals will seriously poison/deactivate the critical catalysts in the process. Cost prohibits carrying out both filtration (soot, additives removal) and re-refining (distilling, isomerisation, hydrocrack, etc.) however the primary hindrance to recycling still remains the collection of fluids as refineries need continuous supply in amounts measured in cisterns, rail tanks.

Occasionally, unused lubricant requires disposal. The best course of action in such situations is to return it to the manufacturer where it can be processed as a part of fresh batches.

Lubricants both fresh and used can cause considerable damage to the environment mainly due to their high potential of serious water pollution. Further the additives typically contained in lubricant can be toxic to flora and fauna. In used fluids the oxidation products can be toxic as well. Lubricant persistence in the environment largely depends upon the base fluid, however if very toxic additives are used they may negatively affect the persistence. Lanolin lubricants are non-toxic making them the environmental alternative which is safe for both users and the environment.

Types of Lubricants

Lubricants are generally composed of a majority of base oil plus a variety of additives to impart desirable characteristics. Although generally lubricants are based on one type of base oil, mixtures of the base oils also are used to meet performance requirements.

Solid Lubricants

Dry lubricants or solid lubricants are materials that, despite being in the solid phase, are able to reduce friction between two surfaces sliding against each other without the need for a liquid oil medium.

The two main dry lubricants are graphite and molybdenum disulfide. They offer lubrication at temperatures higher than liquid and oil-based lubricants operate. Dry lubricants are often used in applications such as locks or dry lubricated bearings. Such materials can operate up to 350 °C (662 °F) in oxidizing environments and even higher in reducing/non-oxidizing environments (molybdenum disulfide up to 1100 °C, 2012 °F). The low-friction characteristics of most dry lubricants are attributed to a layered structure on the molecular level with weak bonding between layers. Such layers are able to slide relative to each other with minimal applied force, thus giving them their low friction properties.

However, a layered crystal structure alone is not necessarily sufficient for lubrication. In fact, there are some solids with non-lamellar structures that function well as dry lubricants in some applications. These include certain soft metals (indium, lead, silver, tin), polytetrafluroethylene, some solid oxides, rare-earth fluorides, and even diamond.

Limited interest has been shown in low friction properties of compacted oxide glaze layers formed at several hundred degrees Celsius in metallic sliding systems. However, practical use is still many years away due to their physically unstable nature.

The four most commonly used solid lubricants are:

- Graphite: Used in air compressors, food industry, railway track joints, brass instrument valves, piano actions, open gear, ball bearings, machine-shop works, etc. It is also very common for lubricating locks, since a liquid lubricant allows particles to get stuck in the lock worsening the problem. It is often used to lubricate the internal moving parts of firearms in sandy environments.

- Molybdenum disulfide (MoS_2): Used in CV joints and space vehicles. Does lubricate in vacuum.
- Hexagonal boron nitride: Used in space vehicles. Also called "white graphite."
- Tungsten disulfide: Similar usage as molybdenum disulfide, but due to the high cost only found in some dry lubricated bearings.

Graphite and molybdenum disulfide are the predominant materials used as dry lubricants.

Structure-function Relationship

The lubricity of many solids is attributable to a lamellar structure. The lamellae orient parallel to the surface in the direction of motion and slide easily over each other resulting in low friction and preventing contact between sliding components even under high loads. Large particles perform best on rough surfaces at low speed, finer particles on smoother surfaces and at higher speeds. These materials may be added in the form of dry powder to liquid lubricants to modify or enhance their properties.

Other components that are useful solid lubricants include boron nitride, polytetrafluorethylene (PTFE), talc, calcium fluoride, cerium fluoride, and tungsten disulfide.

Applications

Solid lubricants are useful for conditions when conventional lubricants are inadequate, such as:

- Reciprocating motion: A typical application is a sliding or reciprocating motion that requires lubrication to minimize wear, as, for example, in gear and chain lubrication. Liquid lubricants will squeeze out while solid lubricants do not escape, preventing fretting, corrosion, and galling.
- Ceramics: Another application is for cases where chemically active lubricant additives have not been found for a particular surface, such as polymers and ceramics.
- High temperature: Graphite and MoS_2 act as lubricants at high temperature and in oxidizing atmosphere environments, where liquid lubricants typically will not survive. A typical application involves fasteners that are easily tightened and unscrewed after a long stay at high temperatures.
- Extreme contact pressures: The lamellar structure orients parallel to the sliding surface, resulting in high bearing-load combined with a low shear stress. Most applications in metal forming that involve plastic deformation use solid lubricants.

Graphite

Graphite is structurally composed of planes of polycyclic carbon atoms that are hexagonal in orientation. The distance of carbon atoms between planes is longer and, therefore, the bonding is weaker.

Graphite is best suited for lubrication in air. Water vapor is a necessary component for graphite lubrication. The adsorption of water reduces the bonding energy between the hexagonal planes of the graphite to a lower level than the adhesion energy between a substrate and the graphite. Because water vapor is a requirement for lubrication, graphite is not effective in vacuum. Because it is electrically conductive, graphite can promote galvanic corrosion. In an oxidative atmosphere, graphite is effective at high temperatures up to 450 °C continuously and can withstand much higher temperature peaks.

Graphite is characterized by two main groups:

- Synthetic graphite is a high temperature sintered product and is characterized by its high purity of carbon (99.5–99.9%). Primary grade synthetic graphite can approach the good lubricity of quality natural graphite.
- Natural graphite is derived from mining. The quality of natural graphite varies as a result of the ore quality and its post-mining processing. The end product is graphite with a content of carbon (high grade graphite 96–98% carbon), sulfur, SiO_2, and ash. The higher the carbon content and the degree of graphitization (high crystalline) the better the lubricity and resistance to oxidation.

For applications where only a minor lubricity is needed and a more thermally insulating coating is required, then amorphous graphite would be chosen (80% carbon).

Molybdenum Disulfide

MoS_2 is mined from some sulfide-rich deposits and refined to achieve a purity suitable for lubricants. Like graphite, MoS_2 has a hexagonal crystal structure with the intrinsic property of easy shear. MoS_2lubrication performance often exceeds that of graphite and is effective in vacuum as well, whereas graphite is not. The temperature limitation of MoS_2 at 400 °C is restricted by oxidation. Particle size and film thickness are important parameters that should be matched to the surface roughness of the substrate. Large particles may result in excessive wear by abrasion caused by impurities in the MoS_2, and small particles may result in accelerated oxidation.

Boron Nitride

Hexagonal boron nitride is a ceramic powder lubricant. The most interesting lubricant feature is its high temperature resistance of 1200 °C service temperature in an oxidizing atmosphere. Furthermore, boron nitride has a high thermal conductivity. (Cubic boron nitride is very hard and used as an abrasive and cutting tool component).

Polytetrafluorethylene

Polytetrafluorethylene (PTFE) is widely used as an additive in lubricating oils and greases. Due to the low surface energy of PTFE, stable unflocculated dispersions of PTFE in oil or water can be produced. Contrary to the other solid lubricants discussed, PTFE does not have a layered structure. The macro molecules of PTFE slip easily along each other, similar to lamellar structures. PTFE shows one of the smallest coefficients of static and dynamic friction, down to 0.04. Operating temperatures are limited to about 260 °C.

Application Methods

Spraying/Dipping/Brushing

Dispersion of solid lubricant as an additive in oil, water, or grease is most commonly used. For parts that are inaccessible for lubrication after assembly, a dry film lubricant can be sprayed. After the solvent evaporates, the coating cures at room temperature to form a solid lubricant. Pastes are grease-like lubricants containing a high percentage of solid lubricants used for assembly and lubrication of highly loaded, slow-moving parts. Black pastes generally contain MoS_2. For high temperatures above 500 °C, pastes are composed on the basis of metal powders to protect metal parts from oxidation necessary to facilitate disassembly of threaded connections and other assemblies.

Free Powders

Dry-powder tumbling is an effective application method. The bonding can be improved by prior phosphating of the substrate. Use of free powders has its limitations, since adhesion of the solid particles to the substrate is usually insufficient to provide any service life in continuous applications. However, to improve running-in conditions or in metal-forming processes, a short duration of the improved slide conditions may suffice.

Anti-friction Coatings

Anti-friction (AF) coatings are "lubricating paints" consisting of fine particles of lubricating pigments, such as molydisulfide, PTFE or graphite, blended with a binder. After application and proper curing, these "slippery" or dry lubricants bond to the metal surface and form a dark gray solid film. Many dry film lubricants contain special rust inhibitors which offer exceptional corrosion protection. Most long-wearing films are of the bonded type but are still restricted to applications where sliding distances are not too long. AF coatings are applied where fretting and galling is a problem (such as splines, universal joints and keyed bearings), where operating pressures exceed the load-bearing capacities of ordinary oils and greases, where smooth running in is desired (piston, camshaft), where clean operation is desired (AF coatings will not collect dirt and debris like greases and oils), and where parts may be stored for long periods.

Composites

Self-lubricating composites: Solid lubricants such as PTFE, graphite, MoS_2 and some other anti-friction and anti-wear additives are often compounded in polymers and all kind of sintered materials. MoS_2, for example, is compounded in materials for sleeve bearings, elastomer O-rings, carbon brushes, etc. Solid lubricants are compounded in plastics to form a "self-lubricating" or "internally lubricated" thermoplastic composite. For example, PTFE particles compounded in the plastic form a PTFE film over the mating surface, resulting in a reduction of friction and wear. MoS_2 compounded in nylon reduces wear, friction and stick-slip. Furthermore, it acts as a nucleating agent effecting in a very fine crystalline structure. The primary use of graphite lubricated thermoplastics is in applications operating in aqueous environments.

Liquid and Oily Lubricants

Animal and vegetable products were certainly man's first lubricants and were used in large quantities. But, because they lack chemical inertness and because lubrication requirements have become more demanding, they have been largely superseded by petroleum products and by synthetic materials. Some organic substances such as lard oil and sperm oil are still in use as additives because of their special lubricating properties.

Petroleum lubricants are predominantly hydrocarbon products extracted from fluids that occur naturally within the Earth. They are used widely as lubricants because they possess a combination of the following desirable properties: (1) availability in suitable viscosities, (2) low volatility, (3) inertness (resistance to deterioration of the lubricant), (4) corrosion protection (resistance to deterioration of the sliding surfaces), and (5) low cost.

Synthetic lubricants generally can be characterized as oily, neutral liquid materials not usually obtained directly from petroleum but having some properties similar to petroleum lubricants. In certain ways they are superior to hydrocarbon products. Synthetics exhibit greater stability of viscosity with temperature changes, resistance to scuffing and oxidation, and fire resistance. Since the properties of synthetics vary considerably, each synthetic lubricant tends to find a special application. A few of the more common classes of synthetics and typical uses of each are shown in table.

Table: Synthetic lubricants and typical applications.

Synthetic lubricant	Typical uses
Dibasic acid esters	Instrument oil, jet turbine lubricant, hydraulic fluid
Phosphate esters	Fire-resistant hydraulic fluid, low-temperature lubricant

Silicones	Damping fluid, low-volatility grease base
Silicate esters	Heat transfer fluid, high-temperature hydraulic fluid
Polyglycol ether compounds	Synthetic engine oil, hydraulic fluids, forming and drawing compounds
Fluorol compounds	Nonflammable fluid, extreme oxidation-resistant lubricant

Another form of oily lubricant is grease, a solid or semisolid substance consisting of a thickening agent in a liquid lubricant. Soaps of aluminum, barium, calcium, lithium, sodium, and strontium are the major thickening agents. Nonsoap thickeners consist of such inorganic compounds as modified clays or fine silicas, or such organic materials as arylureas or phthalocyanine pigments. Lubrication by grease may prove more desirable than lubrication by oil under conditions when (1) less frequent lubricant application is necessary, (2) grease acts as a seal against loss of lubricant and ingress of contaminants, (3) less dripping or splattering of lubricant is called for, or (4) less sensitivity to inaccuracies in the mating parts is needed.

Semi-solid Lubricants/Grease

Grease is a semisolid lubricant. Grease generally consists of a soap emulsified with mineral or vegetable oil. The characteristic feature of greases is that they possess a high initial viscosity, which upon the application of shear, drops to give the effect of an oil-lubricated bearing of approximately the same viscosity as the base oil used in the grease. This change in viscosity is called shear thinning. Grease is sometimes used to describe lubricating materials that are simply soft solids or high viscosity liquids, but these materials do not exhibit the shear-thinning properties characteristic of the classical grease. For example, petroleum jellies such as Vaseline are not generally classified as greases.

Greases are applied to mechanisms that can be lubricated only infrequently and where a lubricating oil would not stay in position. They also act as sealants to prevent ingress of water and incompressible materials. Grease-lubricated bearings have greater frictional characteristics because of their high viscosity.

Properties

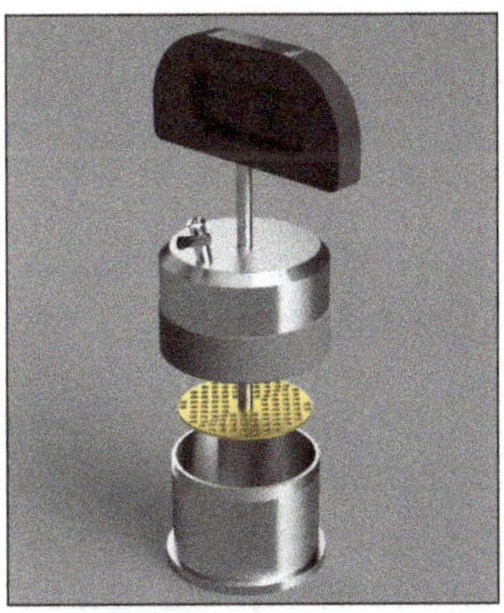

Grease Worker, 3D-Model.

A true grease consists of an oil and/or other fluid lubricant that is mixed with a thickener, typically a soap, to form a solid or semisolid. Greases are a type of shear-thinning or pseudo-plastic fluid, which means that the viscosity of the fluid is reduced under shear. After sufficient force to shear the grease has been applied, the viscosity drops and approaches that of the base lubricant, such as the mineral oil. This sudden drop in shear force means that grease is considered a plastic fluid, and the reduction of shear force with time makes it thixotropic. It is often applied using a grease gun, which applies the grease to the part being lubricated under pressure, forcing the solid grease into the spaces in the part.

Thickeners

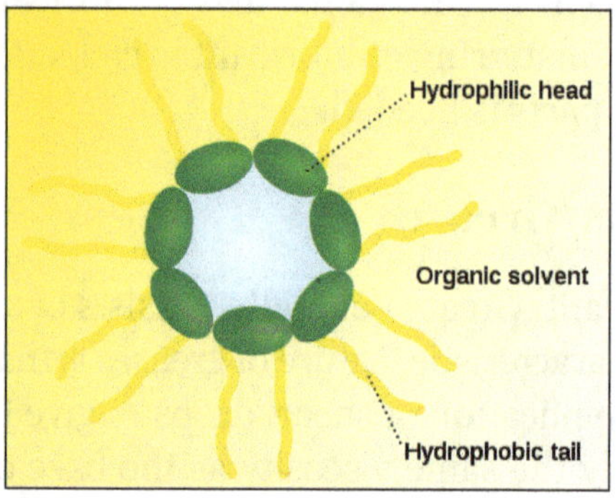

An inverse micelle formed when a soap is dispersed in an oil. This structure is broken reversibly upon shearing the grease.

Soaps are the most common emulsifying agent used, and the selection of the type of soap is determined by the application. Soaps include calcium stearate, sodium stearate, lithium stearate, as well as mixtures of these components. Fatty acids derivatives other than stearates are also used, especially lithium 12-hydroxystearate. The nature of the soaps influences the temperature resistance (relating to the viscosity), water resistance, and chemical stability of the resulting grease.

Powdered Solid Greases

Powdered solids may also be used as thickeners, especially as clays, which are used in some inexpensive, low performance greases. Fatty oil-based greases have also been prepared with other thickeners, such as tar, graphite, or mica, which also increase the durability of the grease.

Engineering Assessment and Analysis

Lithium-based greases are the most commonly used; sodium and lithium-based greases have higher melting point (dropping point) than calcium-based greases but are not resistant to the action of water. Lithium-based grease has a dropping point at 190 to 220 °C (350 to 400 °F). However the maximum usable temperature for lithium-based grease is 120 °C.

The amount of grease in a sample can be determined in a laboratory by extraction with a solvent followed by e.g. gravimetric determination.

Additives

Gear greases consist of rosin oil, condensed with lime and stirred with mineral oil, with some percentage of water. Special-purpose greases contain glycerol and sorbitan esters. They are used, for example, in low-temperature conditions. Some greases are labeled "EP", which indicates "extreme pressure". Under high pressure or shock loading, normal grease can be compressed to the extent that the greased parts come into physical contact, causing friction and wear.

EP grease contains solid lubricants, usually graphite and/or molybdenum disulfide, to provide protection under heavy loadings. The solid lubricants bond to the surface of the metal, and prevent metal-to-metal contact and the resulting friction and wear when the lubricant film gets too thin.

Solid additives such as copper or ceramic powder are added to some greases for static high pressure and/or high temperature applications, or where corrosion could prevent dis-assembly of components later in their service life. These compounds are working as a release agent. Solid additives cannot be used in bearings because of tight tolerances. Solid additives will cause increased wear in bearings.

Classification and Standards

Red wheel bearing grease for automotive applications.

Jointly developed by ASTM International, the National Lubricating Grease Institute (NLGI) and SAE International, standard ASTM D4950 "standard classification and specification for automotive service greases" was first published in 1989 by ASTM International. It categorizes greases suitable for the lubrication of chassis components and wheel bearings of vehicles, based on performance requirements, using codes adopted from the NLGI's "chassis and wheel bearing service classification system":

- LA and LB: Chassis lubricants (suitability up to mild and severe duty respectively).

- GA, GB and GC: Wheel-bearings (suitability up to mild, moderate and severe duty respectively).

A given performance category may include greases of different consistencies. The measure of the consistency of grease is commonly expressed by its NLGI consistency number.

The main elements of standard ATSM D4950 and NLGI's consistency classification are reproduced and described in standard SAE J310 "automotive lubricating greases" published by SAE International.

Standard ISO 6743-9 "lubricants, industrial oils and related products (class L) – classification – part 9: family X (greases)", first released in 1987 by the International Organization for Standardization, establishes a detailed classification of greases used for the lubrication of equipment, components of machines, vehicles, etc. It assigns a single multi-part code to each grease based on its operational properties (including temperature range, effects of water, load, etc.) and its NLGI consistency number.

Silicone Grease

Silicone grease, sometimes called dielectric grease, is a waterproof grease made by combining a silicone oil with a thickener. Most commonly, the silicone oil is polydimethylsiloxane (PDMS) and the thickener is amorphous fumed silica. Using this formulation, silicone grease is a translucent white viscous paste, with exact properties dependent on the type and proportion of the components. More specialized silicone greases are made from fluorinated silicones or, for low-temperature applications, PDMS containing some phenyl substituents in place of methyl groups. Other thickeners may be used, including stearates and powdered polytetrafluorethylene (PTFE).

Use in Industry

Silicone grease is commonly used for lubricating and preserving rubber parts, such as O-rings. Additionally, silicone grease does not swell or soften the rubber, which can be a problem with hydrocarbon-based greases. It functions well as a corrosion inhibitor and lubricant for purposes that require a thicker lubricant.

Silicone grease is soluble in organic solvents such as toluene, xylene, mineral spirits, and chlorinated hydrocarbons. It is insoluble in methanol, ethanol, and water.

Thermal grease often consists of a silicone-grease base, along with added thermally conductive fillers. It is used for heat-transfer abilities, rather than friction reduction.

Special versions of silicone grease are also used widely by the plumbing industry in faucets and seals, as well as in dental equipment. These special versions are formulated using components not known to be an ingestion hazard. Electrical utilities use silicone grease to lubricate separable elbows on lines that must endure high temperatures. Silicone greases generally have an operating temperature range of approximately −40 to 200 °C (−40 to 392 °F) with some high-temperature versions extending this range slightly.

Use in the Chemical Laboratory

Silicone grease is widely used as a temporary sealant and a lubricant for interconnecting ground glass joints, as is typically used in laboratory glassware. Although silicones are normally assumed to be chemically inert, several historically significant compounds have resulted from unintended reactions with silicones. The first salts of crown ethers $(OSi(CH_3)_2)_n$ (n = 6, 7) were produced by reactions of organolithium and organopotassium compounds with silicone greases or the serendipitous reaction of stannanetriol with silicone grease to afford a cage-like compound having three Sn–O–Si–O–Sn linkages in the molecule.

Lubrication of an apparatus with silicone grease may result in the reaction mixture being contaminated with the grease. The impurity may be carried through purification by chromatography in undesirable amounts. In NMR spectroscopy, the methyl groups in polydimethylsiloxane display ^{1}H and ^{13}C chemical shifts similar to trimethylsilane (TMS), the reference compound for those forms of NMR spectroscopy. As with TMS, the signal is a singlet. In ^{1}H NMR, silicone grease appears at a singlet at $\delta = 0.07$ ppm in $CDCl_3$, 0.09 in CD_3CN, 0.29 in C_6D_6, and −0.06 ppm in $(CD_3)_2SO$. In ^{13}C NMR, it appears at $\delta = 1.19$ ppm in $CDCl_3$ and 1.38 ppm in C_6D_6. Tables of impurities commonly found in NMR spectroscopy have been prepared, and such tables include silicone grease.

Consumer Uses

Silicone-based lubricants are often used by consumers in applications where other common consumer lubricants, such as petroleum jelly, would damage certain products, such as latex rubber and gaskets on dry-suits. It can be used to lubricate fountain pen filling mechanisms and threads. It is used to seal and preserve O-rings in flashlights, plumbing, waterproof watches, and air rifles. Silicone grease is widely used to lubricate threads of water-submersible flashlights used for diving and spearfishing. This grease improves water resistance of the flashlights and protects threads from wearing out. Silicone grease is used with waterproof devices, as it has a very thick body and doesn't dissolve in water as most spirits and other liquids would.

Various household uses include lubricating door hinges, shower heads, threads on bolts, garden-hose threads or any thread or mechanism that can be lubricated.

As a Sealant around Electrical Contacts

Silicone greases are electrically insulating and are often applied to electrical connectors, particularly those containing rubber gaskets, as a means of sealing and protecting the connector. In this context they are often referred to as dielectric grease.

A common use of this type is in the high-voltage connection associated with gasoline-engine spark plugs, where grease is applied to the rubber boot of the plug wire to help it slide onto the ceramic insulator of the plug, to seal the rubber boot, and to prevent the rubber's adhesion to the ceramic. Such greases are formulated to withstand the high temperature generally associated with the areas in which spark plugs are located, and can be applied to contacts as well (because the contact pressure is sufficient to penetrate the grease film). Doing so on such high-pressure contact surfaces between different metals has the further advantage of sealing the contact area against electrolytes that might cause rapid deterioration of the metals by galvanic corrosion.

Silicone grease can decompose to form an insulating layer at or next to switch contacts that experience arcing, and contamination can cause the contacts to prematurely fail.

Cork Grease

Cork grease is a grease for woodwind and reed instruments such as saxophones, clarinets, bassoons, and oboes. It is used on the corks to make putting the instrument together possible without damaging the cork and therefore possibly damaging the instrument's barrel. Cork grease also acts as a preservative, keeping the wooden cork moist and thick, in turn ensuring a good seal between parts of the instrument so that no air may leak through joints upon playing. Cork grease can help woodwind players adjust their instruments' tuning pieces (e.g. barrels, necks, bocals, staples) in respect to their pitch.

Cork grease is made with ingredients such as elm extract, organic sunflower oil, coconut oil and hemp seed oil. In past it was made from animal fat. It is not toxic to humans.

Lithium 12-hydroxystearate

Lithium 12-hydroxystearate is a chemical compound classified as a lithium soap. In chemistry, "soap" refers to salts of fatty acids. Lithium 12-hydroxystearate is a white solid. Lithium soaps are key component of many lubricating greases.

Lithium 12-hydroxystearate exhibits high oxidation stability and a dropping point up to around 200 °C. Most greases used today in motor vehicles, aircraft, and heavy machinery contain lithium stearates, mainly lithium 12-hydroxystearate. Greases can be made with the addition of several different metallic soaps. Some greases are prepared from sodium, barium, lithium, and calcium soaps. Lithium soap greases are preferred for their water resistance, and their oxidative and mechanical stability. Depending on the grease, they also have good performance at high or low temperatures, but not both.

Production

To produce lithium 12-hydroxystearate, lithium hydroxide and the fatty acid are combined in an aqueous medium. With vigorous stirring, dilute monohydrate lithium hydroxide is gradually added to a dispersion of the fatty acid in water heated to slightly below boiling. Since these lithium soaps are difficult to filter, they are collected by spray drying. For applications, lithium 12-hydroxystearate is usually dispersed in synthetic oils such as silicone oil and ester oil. The synthetic oils are preferred for their greater stability and ability to perform at extreme temperatures.

The 12-hydroxystearic acid is prepared by the hydrogenation of castor oil. After primary reaction of the saturation of most of the double bonds, dehydration and reduction of the hydroxyl group leads to the stearic acid. Hydrogenated castor oil results in a mixture of 12-hydroxystearic acid and stearic acid.

Base Oil

Base oils are used to manufacture products including lubricating greases, motor oil and metal processing fluids. Different products require different compositions and properties in the oil. One of the most important factors is the liquid's viscosity at various temperatures. Whether or not a crude oil is suitable to be made into a base oil is determined by the concentration of base oil molecules as well as how easily these can be extracted.

Base oil is produced by means of refining crude oil. This means that crude oil is heated in order that various distillates can be separated from one another. During the heating process, light and heavy hydrocarbons are separated – the light ones can be refined to make petrol and other fuels, while the heavier ones are suitable for bitumen and base oils.

There are large numbers of crude oils all around the world that are used to produce base oils. The most common one is a type of paraffinic crude oil, although there are also naphthenic crude oils that create products with better solubility and very good properties at low temperatures. By using hydrogenation technology, in which sulfur and aromatics are removed using hydrogen under high pressure, you can obtain extremely pure base oils, which are suitable when quality requirements are particularly stringent.

Chemical substances – additives – are added to the base oil in order to meet the quality requirements for the end products in terms of, for example, friction and cleaning properties. Certain types of motor oils contain more than twenty percent additives.

Official Classifications

In 1993, the American Petroleum Institute (API), categorized base oils into five main groups. This breakdown is based on the refining method and the base oil's properties in terms of, among other things, viscosity and the proportion of saturates and sulfur content.

- Group I: Originating in the 1930s, the least refined type which is produced by Solvent Refining. It usually consists of conventional petroleum base oils. An improvement to the refining process in the 1960s called hydro-treating made this base oil more stable, less reactive, and longer lasting than the earlier base oils.

API defines group I as "base stocks contain less than 90 percent saturates and/or greater than 0.03 percent sulfur and have a viscosity index greater than or equal to 80 and less than 120".

- Group II: Originating in 1971, a better grade of petroleum base oil, which may be partially produced by Hydrocracking. All impurities will be removed from the oil leading to clearer color.

  API defines group II as "base stocks contain greater than or equal to 90 percent saturates and less than or equal to 0.03 percent sulfur and have a viscosity index greater than or equal to 80 and less than 120".

- Group III: Originating in 1993, the best grade of petroleum base oil, since they are fully produced by Hydrocracking, Hydroisomerization, and Hydrotreating, which make these oils purer.

  API defines group III as "base stocks contain greater than or equal to 90 percent saturates and less than or equal to 0.03 percent sulfur and have a viscosity index greater than or equal to 120".

  This group may be described as Synthetic Technology oils or Hydro-Cracked Synthetic oil. However, some oil companies may call their products under this group as synthetic oil.

- Group IV: Originating in 1974, consists of synthetic oils made of Poly-alpha-olefins (PAO). Group IV base oils have a viscosity index range of 125 - 200.

  Poly-alpha-olefins (PAO) oils have a higher oxidative stability in extreme temperatures, and also have exceptionally low pour points, which makes them much more suitable for use in very cold weather (as found in northern Europe), as well as in very hot weather (as in Middle East).

- Group V: They include, among others, naphthenic oils and polyesters.

Unofficial Classifications

Unofficial base oil classifications are not recognized by the American Petroleum Institute (API), however, they are widely used and marketed for motor oils and automatic transmission fluids.

Group II+

Originating in the 1990s, a more refined grade of petroleum Group III base oil, produced by Hydrotreating. Group II+ base oils have a high viscosity index at the higher end of the API Group II range. The viscosity index is 110-115 minimum.

Group III+

Originating in 2015, produced by a gas to liquids (GTL) process. Group III+ base oils have a Very High Viscosity Index (VHVI) at the higher end of the API Group III range. The viscosity index is 130-140 minimum.

Group VI

Consists of synthetic oils made of Poly-internal-olefins (PIO).

Poly-internal-olefins (PIO) oils are similar to Poly-alpha-olefins (PAO), but use different chemicals in the synthesis process to obtain an even higher viscosity index (VI).

Synthetic Oil

A sample of synthetic motor oil.

Synthetic oil is a lubricant consisting of chemical compounds that are artificially made. Synthetic lubricants can be manufactured using chemically modified petroleum components rather than whole crude oil, but can also be synthesized from other raw materials. The base material, however, is still overwhelmingly crude oil that is distilled and then modified physically and chemically. The actual synthesis process and composition of additives is generally a commercial trade secret and will vary among producers.

Synthetic oil is used as a substitute for petroleum-refined oils when operating in extreme temperature. Aircraft jet engines, for example, require the use of synthetic oils, whereas aircraft piston engines do not. Synthetic oils are also used in metal stamping to provide environmental and other benefits when compared to conventional petroleum and animal-fat based products. These products are also referred to as "non-oil" or "oil free".

Types

Full

Some "synthetic" oil is made from Group III base stock, some from Group IV. Some from a blend of the two. Mobil sued Castrol and Castrol prevailed in showing that their Group III base stock oil was changed enough that it qualified as full synthetic. Since then API has removed all references to Synthetic in their documentation regarding standards. "Full synthetic" is a marketing term and is not a measurable quality.

Group IV: PAO

Poly-alpha-olefin (poly-α-olefin, PAO) is a polymer made by polymerizing an alpha-olefin. They are designated at API Group IV and are a 100% synthetic chemical compound. It is a specific type of olefin (organic) that is used as a base stock in the production of some synthetic lubricants. An alpha-olefin (or α-olefin) is an alkene where the carbon-carbon double bond starts at the α-carbon atom, i.e. the double bond is between 1 and 2 carbons in the molecule.

Group V: Other Synthetics

Group V base oils are defined by API as any other type of oil other than mineral oils or PAO lubricants.

Esters are the most famous synthetics in Group V, which are 100% synthetic chemical compounds consisting of a carbonyl adjacent to an ether linkage. They are derived by reacting an oxoacid with a hydroxyl compound such as an alcohol or phenol. Esters are usually derived from an inorganic acid or organic acid in which at least one -OH (hydroxyl) group is replaced by an -O-alkyl (alkoxy) group, most commonly from carboxylic acids and alcohols. That is to say, esters are formed by condensing an acid with an alcohol.

Many chemically different "esters" due to their usually excellent lubricity are used for various reasons as either "additives" or "base stocks" for lubricants.

Polyalkylene Glycol Synthetic (PAG) Oil

Industrial Polyalkylene Glycols

The terms polyalkylene glycol and polyglycol are used interchangeably. Synthetic lubricants are about 4% of the lubricants market. Polyalkylene glycols (PAG) are about 24% of the synthetic lubricants market.

Ethylene is the basic raw material used to make the synthetic lubricant polyglycols oils. When ethylene and propylene react with oxygen, we obtain ethylene oxide (EO) and propylene oxide (PO), from which the polyalkylene glycols are produced by means of polymerization. PAGs are usually made by combining ethylene oxide (EO) and/or propylene oxide (PO) with an alcohol or water.

The mixing ratio between EO and PO, plus the oxygen bonded in the chemical structure, crucially affect the behavior of polyglycols. The gear industry predominantly uses polyglycols with an EO/PO ratio of 50:50 to 60:40, which exhibit very similar behavior. The polyglycols featuring this composition are also generally referred to as water soluble polyglycols. Polyalkylene Glycol Base oils (PAGs) are formed by reacting an alcohol with one or more alkylene oxides: Propylene oxide provides water insolubility and Ethylene oxide provides water solubility.

Polyalkylene Glycol Properties

PAGs offer properties that include; high lubricity, low traction properties, high viscosity index, controlled quenching speeds, good temperature stability and low wear. They are available in both water soluble and insoluble forms, and in a wide range of viscosity grades.

Polyalkylene Glycol Uses

PAGS are commonly used in quenching fluids, metalworking fluids, gear oils, chain oils, food-grade lubricants and as lubricants in HFC type hydraulics and gas compressor equipment. PAG lubricants are used by the two largest U.S. air compressor OEMs in rotary screw air compressors.

PAG Advantages

- PAGs are available with a wide range viscosities for a variety of uses. Some PAGs properties such as water-soluble polymers, are not commonly provided by other synthetic lubricants, such as polyalphaolefins (PAOs).
- PAGs prevent sludge and varnish from developing at high temperatures. PAGs have viscosity indexes that are higher than PAOs.
- In large gears, polyalkylene glycol lubricant yielded lower friction than polyalphaolefin lubricant.
- PAGs can be highly biodegradable, particularly the water soluble PAGs.

PAG Disadvantages

PAGs are not compatible with mineral oils, most seals, paints, varnishes.

PAG Seal Compatibility

PAG is usually compatible with FKM fuoroelastomer materials and VMQ (vinyl methyl silicone) Silicone rubber.

PAG acts as a solvent and dissolves and removes the mineral grease which causes slower motions, air leaks, and can stop 4-way valves from operating. Natural rubber, Buna-N, and most regular seals are incompatible with PAG oils, especially seals coated in mineral grease. PAG oils can cause seals to shrink or swell, thus causing severe leakage or seizure of the seal. Pneumatic air cylinders and 4-ways valves commonly use Buna-N rubber seals that coated in mineral grease.

Semi-synthetic Oil

Semi-synthetic oils (also called "synthetic blends") are a mixture of mineral oil and synthetic oil, which are engineered to have many of the benefits of full synthetic oil without the cost. Motul introduced the first semi-synthetic motor oil in 1966.

Lubricants that have synthetic base stocks even lower than 30% but with high-performance additives consisting of esters can also be considered synthetic lubricants. In general, the ratio of the synthetic base stock is used to define commodity codes among the customs declarations for tax purposes.

Other Base Stocks help Semi-synthetic Lubricants

API Group II- and API Group III-type base stocks help to formulate more economic-type semi-synthetic lubricants. API Group I-, II-, II+-, and III-type mineral-base oil stocks are widely used in combination with additive packages, performance packages, and ester and/or API Group IV poly-alpha-olefins in order to formulate semi-synthetic-based lubricants. API Group III base oils are sometimes considered fully synthetic, but they are still classified as highest-top-level mineral-base stocks. A synthetic or synthesized material is one that is produced by combining or building individual units into a unified entity. Synthetic base stocks as described above are man-made and tailored to have a controlled molecular structure with predictable properties, unlike mineral base oils, which are complex mixtures of naturally occurring hydrocarbons and paraffins.

Performance

The advantages of using synthetic motor oils includes better low- and high-temperature viscosity performance at service temperature extremes better (higher) Viscosity Index (VI) and chemical and shear stability. This also helps in decreasing the loss due to evaporation. It serves resistance to oxidation, thermal breakdown, and oil sludge problems and extended drain intervals, with the environmental benefit of less used oil waste generated. It provides better lubrication during the time when extreme cold weather starts The use of synthetic oils promise possibly a longer engine life with superior protection against "ash" and other deposit formation in engine hot spots (in particular in turbochargers and superchargers) for less oil burn-off and reduced chances of damaging oil passageway clogging. The performance of automobiles is improved as net increase in horsepower and torque due to less internal drag on engine. Moreover, it helps in improved fuel efficiency - from 1.8% to up to 5% has been documented in fleet tests. Researches suggest that synthetics performed about 47% better than regular oil.

However, synthetic motor oils are substantially more expensive (per volume) than mineral oils and have potential decomposition problems in certain chemical environments (predominantly in industrial use).

Gaseous Lubricants

Lubrication with a gas is analogous in many respects to lubrication with a liquid, since the same principles of fluid-film lubrication apply. Although both gases and liquids are viscous fluids, they differ in two important particulars. The viscosity of gases is much lower and the compressibility much greater than for liquids. Film thicknesses and load capacities therefore are much lower with a gas such as air. In equipment that handles gases of various kinds, it is often desirable to lubricate the sliding surfaces with gas in order to simplify the apparatus and reduce contamination to and from the lubricant. The list of gases used in this manner is extensive and includes air, steam, industrial gases, and liquid-metal vapours.

With so many types of materials capable of acting as lubricants under certain conditions, coverage of the properties of all of them is impractical. Mention is made only of those properties usually considered characteristic of commercially significant fluid lubricants.

Lubrication Mechanism

Fluid-lubricated systems are designed so that the applied load is partially or completely carried by hydrodynamic or hydrostatic pressure, which reduces solid body interactions (and consequently friction and wear). Depending on the degree of surface separation, different lubrication regimes can be distinguished.

Adequate lubrication allows smooth, continuous operation of machine elements, reduces the rate of wear, and prevents excessive stresses or seizures at bearings. When lubrication breaks down, components can rub destructively against each other, causing heat, local welding, destructive damage and failure.

Lubrication of a ship's steam enginecrankshaft. The two bottles of lubricant are attached to the piston and move while the engine is operating.

Fluid-lubricated Systems

As the load increases on the contacting surfaces, three distinct situations can be observed with respect to the mode of lubrication, which are called lubrication regimes:

- Fluid film lubrication is the lubrication regime in which, through viscous forces, the load is fully supported by the lubricant within the space or gap between the parts in motion relative to one another object (the lubricated conjunction) and solid–solid contact is avoided.
    - In hydrostatic lubrication, external pressure is applied to the lubricant in the bearing to maintain the fluid lubricant film where it would otherwise be squeezed out.
    - In hydrodynamic lubrication the motion of the contacting surfaces as well as the design of the bearing pump lubricant around the bearing to maintain the lubricating film. This design of bearing may wear when started, stopped or reversed, as the lubricant film breaks down. The basis of the hydrodynamic theory of lubrication is the Reynolds equation. The governing equations of the hydrodynamic theory of lubrication and some analytical solutions can be found in the reference.
- Elastohydrodynamic lubrication: Mostly for nonconforming surfaces or higher load conditions, the bodies suffer elastic strains at the contact. Such strain creates a load-bearing area, which provides an almost parallel gap for the fluid to flow through. Much as in hydrodynamic lubrication, the motion of the contacting bodies generates a flow induced pressure, which acts as the bearing force over the contact area. In such high pressure regimes, the viscosity of the fluid may rise considerably. At full film elastohydrodynamic lubrication the generated lubricant film completely separates the surfaces. Contact between

raised solid features, or asperities, can occur, leading to a mixed-lubrication or boundary lubrication regime. In addition to Reynolds equation, elastohydrodynamic theory considers the elastic deflection equation, since in this regime elastic deformation of the surfaces contributes significantly to the lubricant film thickness.

- Boundary lubrication (also called boundary film lubrication): The hydrodynamic effects are negligible. The bodies come into closer contact at their asperities; the heat developed by the local pressures causes a condition which is called stick-slip, and some asperities break off. At the elevated temperature and pressure conditions, chemically reactive constituents of the lubricant react with the contact surface, forming a highly resistant tenacious layer or film on the moving solid surfaces (boundary film) which is capable of supporting the load and major wear or breakdown is avoided. Boundary lubrication is also defined as that regime in which the load is carried by the surface asperities rather than by the lubricant.
- Mixed lubrication: This regime is in between the full film elastohydrodynamic and boundary lubrication regimes. The generated lubricant film is not enough to separate the bodies completely, but hydrodynamic effects are considerable.

Besides supporting the load the lubricant may have to perform other functions as well, for instance it may cool the contact areas and remove wear products. While carrying out these functions the lubricant is constantly replaced from the contact areas either by the relative movement (hydrodynamics) or by externally induced forces.

Lubrication is required for correct operation of mechanical systems such as pistons, pumps, cams, bearings, turbines, cutting tools etc. where without lubrication the pressure between the surfaces in close proximity would generate enough heat for rapid surface damage which in a coarsened condition may literally weld the surfaces together, causing seizure.

In some applications, such as piston engines, the film between the piston and the cylinder wall also seals the combustion chamber, preventing combustion gases from escaping into the crankcase.

Properties of Lubricants

A good lubricant generally possesses the following characteristics:

- A high boiling point and low freezing point (in order to stay liquid within a wide range of temperature).
- A high viscosity index.

- Thermal stability.
- Hydraulic stability.
- Demulsibility.
- Corrosion prevention.
- A high resistance to oxidation.

Formulation

Typically lubricants contain 90% base oil (most often petroleum fractions, called mineral oils) and less than 10% additives. Vegetable oils or synthetic liquids such as hydrogenated polyolefins, esters, silicones, fluorocarbons and many others are sometimes used as base oils. Additives deliver reduced friction and wear, increased viscosity, improved viscosity index, resistance to corrosion and oxidation, aging or contamination, etc.

Non-liquid lubricants include powders (dry graphite, PTFE, molybdenum disulphide, tungsten disulphide, etc), PTFE tape used in plumbing, air cushion and others. Dry lubricants such as graphite, molybdenum disulphide and tungsten disulphide also offer lubrication at temperatures (up to 350 °C) higher than liquid and oil-based lubricants are able to operate. Limited interest has been shown in low friction properties of compacted oxide glaze layers formed at several hundred degrees Celsius in metallic sliding systems, however, practical use is still many years away due to their physically unstable nature.

Additives

A large number of additives are used to impart performance characteristics to the lubricants. Modern automotive lubricants contain as many as ten additives, comprising up to 20% of the lubricant, the main families of additives are:

- Pour point depressants are compounds that prevent crystallization of waxes. Long chain alkylbenzenes adhere to small crystallites of wax, preventing crystal growth.
- Anti-foaming agents are typically silicone compounds which increase surface tension in order to discourage foam formation.
- Viscosity index improvers (VIIs) are compounds that allow lubricants to remain viscous at higher temperatures. Typical VIIs are polyacrylates and butadiene.
- Antioxidants suppress the rate of oxidative degradation of the hydrocarbon molecules within the lubricant. At low temperatures, free radical inhibitors such as hindered phenols are used, e.g. butylated hydroxytoluene. At temperatures

>90 °C, where the metals catalyze the oxidation process, dithiophosphates are more useful. In the latter application the additives are called metal deactivators.

- Detergents ensure the cleanliness of engine components by preventing the formation of deposits on contact surfaces at high temperatures.
- Corrosion inhibitors (rust inhibitors) are usually alkaline materials, such as alkylsulfonate salts, that absorb acids that would corrode metal parts.
- Anti-wear additives form protective 'tribofilms' on metal parts, suppressing wear. They come in two classes depending on the strength with which they bind to the surface. Popular examples include phosphate esters and zinc dithiophosphates.
- Extreme pressure (anti-scuffing) additives form protective films on sliding metal parts. These agents are often sulfur compounds, such as dithiophosphates.
- Friction modifiers reduce friction and wear, particularly in the boundary lubrication regime where surfaces come into direct contact.

Lubricant Additives

Lubricant additives are organic or inorganic compounds dissolved or suspended as solids in oil. They typically range between 0.1 to 30 percent of the oil volume, depending on the machine.

Additives have three basic roles:

- Enhance existing base oil properties with antioxidants, corrosion inhibitors, anti-foam agents and demulsifying agents.
- Suppress undesirable base oil properties with pour-point depressants and viscosity index (VI) improvers.
- Impart new properties to base oils with extreme pressure (EP) additives, detergents, metal deactivators and tackiness agents.

Polar Additives

Additive polarity is defined as the natural directional attraction of additive molecules to other polar materials in contact with oil. In simple terms, it is anything that water dissolves or dissolves into water.

A sponge, a metal surface, dirt, water and wood pulp are all polar. Things that are not polar include wax, Teflon, mineral base stock, a duck's back and water repellents.

It's important to note that additives are also sacrificial. Once they are gone, they're gone. Think about the environment you work in, the products you produce and the types of contaminants that are around you daily. If you are allowing into your system contaminants that additives are attracted to, such as dirt, silica and water, the additives will cling to the contaminants and settle to the bottom or will be filtered out and deplete your additive package.

Enhance Existing Base Oil Properties	Suppress Undesirable Base Oil Properties	Impart New Properties to Base Oils
Antioxidants Corrosion Inhibitors Anti-foam Agents Demulsifying Agents	Pour Point Depressants VI Improvers	EP Additives Detergents Metal Deactivators Tackiness Agents

Polar Mechanisms

There are a few polar mechanisms such as particle enveloping, water emulsifying and metal wetting that are worthy of discussion. Particle enveloping means that the additive will cling to the particle surface and envelop it. These additives are metal deactivators, detergents and dispersants. They are used to peptize (disperse) soot particles for the purpose of preventing agglomeration, settling and deposits, especially at low to moderate temperatures.

You generally will see this in an engine. It offers a good reason to repair and eliminate any issues as soon as they are detected through an appropriate oil analysis test slate. When using oil additives, more is not always better. As more additive is blended into the oil, sometimes there isn't any more benefit gained, and at times the performance actually deteriorates. In other cases, the performance of the additive doesn't improve, but the duration of service does improve.

In addition, increasing the percentage of a certain additive may improve one property of an oil while at the same time degrade another. When the specified concentrations of additives become unbalanced, overall oil quality can also be affected.

Some additives compete with each other for the same space on a metal surface. If a high concentration of an anti-wear agent is added to the oil, the corrosion inhibitor may become less effective. The result may be an increase in corrosion-related problems. Water emulsifying occurs when the additive polar head clings to a micro-droplet of moisture. These types of additives are emulsifying agents. Consider this the next time you observe water in a reservoir.

While it is important to remove the water, determine where the water entered the system and repair it using a root-cause maintenance approach, you must also keep in mind that the additive package has been affected. In lubrication terms, this is known as additive depletion. A proper oil analysis report can determine the health of the additives remaining in the lubricant.

Machine	Common Additives used	Percent of oil volume
Engines	Antioxidant, corrosion inhibitor, detergent/dispersant, anti-wear, anti-foam, alkalinity improver	10-30%
Steam turbines, compressors	Antioxidant, corrosion inhibitor, demulsifier, anti-foam	0.5-5%
Gears (spiral, bevel or hypoid)	Anti-wear, antioxidant, anti-foam, sometimes corrosion inhibitor, extreme pressure	1-10%
Gears (worm)	Extreme pressure, antioxidant, corrosion inhibitor, fatty acids	3-10%
Hydraulic systems	Antioxidant, anti-wear, anti-foam, corrosion inhibitor, pour-point depressant, viscosity index improver	2-10%

Metal wetting is when additives anchor to metal surfaces, which is what they are supposed to do. They attach to the interior of the gear casing, gear teeth, bearings, shafts, etc.

Additives that perform this function are rust inhibitors, anti-wear (AW) and EP additives, oiliness agents and corrosion inhibitors. AW additives work specifically to protect metal surfaces during boundary conditions. They form a ductile, ash-like film at moderate to high contact temperatures (150 to 230 degrees F). Under boundary conditions, AW film shears instead of surface material.

One common anti-wear additive is zinc dialkyldithiophosphate (ZDDP). It reduces the risk of metal-to-metal contact, which can lead to increased heat, result in oxidation and negatively affect the film strength. Whether they are enhancing, suppressing or imparting new properties to the base oil, additives play an important role in the lubrication of machinery. Remember, when the additives are gone, they're gone, so don't forget to check your additive package.

Types of Lubricant Additives

There are many types of chemical additives mixed into base oils to enhance the properties of the base oil, to suppress some undesirable properties of the base oil and possibly to impart some new properties:

- Additives typically make up about 0.1 to 30 percent of the finished lubricating oil, depending upon the target application of the lubricant.
- Lubricant additives are expensive chemicals, and creating the proper mix or formulation of additives is a very complicated science. It is the choice of additives that differentiates a turbine (R&O) oil from a hydraulic oil, a gear oil and an engine oil.

- Many lubricant additives are available, and they are selected for use based upon their ability to perform their intended function. They are also chosen for their ability to mix easily with the selected base oils, to be compatible with other additives in the formulation and to be cost effective.
- Some additives perform their function within the body of the oil (e.g., anti-oxidants), while others do their work on the surface of the metal (e.g., anti-wear additives and rust inhibitors).

Conventional Lubricant Additives

These include the following general types of additives:

Anti-oxidants

Oxidation is the general attack of the weakest components of the base oil by oxygen in the air. It occurs at all temperatures all of the time but is accelerated at higher temperatures and by the presence of water, wear metals and other contaminants. It ultimately causes acids (which produce corrosion) and sludge (which results in surface deposits and viscosity to increase) to form. Oxidation inhibitors, as they are also called, are used to extend the operating life of the oil.

They are sacrificial additives that are consumed while performing their duty of delaying the onset of oxidation, thus protecting the base oil. They are present in almost every lubricating oil and grease.

Rust and Corrosion Inhibitors

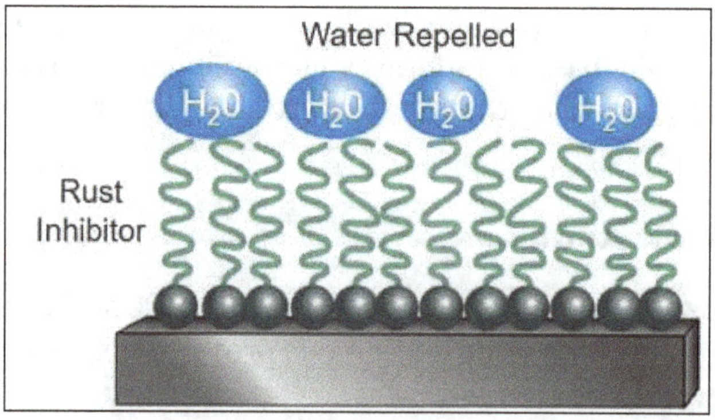

Iron or steel surface.

These additives reduce or eliminate internal rust and corrosion by neutralizing acids and forming a chemical protective barrier to repel moisture from metal surfaces.

Some of these inhibitors are specific to protecting certain metals. Therefore, an oil may contain several corrosion inhibitors. Again, they are common in almost every oil and grease. Metal deactivators are another form of corrosion inhibitor.

Viscosity Index Improvers

- Viscosity index improvers are very large polymer additives that partially prevent the oil from thinning out (losing viscosity) as the temperature increases. These additives are used extensively when blending multi-grade engine oils such as SAE 5W-30 or SAE 15W-40.
- They are also responsible for better oil flow at low temperatures, resulting in reduction in wear and improved fuel economy. In addition, VI improvers are used to achieve high-VI hydraulic and gear oils for improved start-up and lubrication at low temperatures.
- To visualize how a VI-improver additive functions, think of the VI improver as an octopus or coil spring that stays coiled up in a ball at low temperatures and has very little effect on the oil viscosity.
- Then, as the temperature rises, the additive (or octopus) expands or extends its arms (making it larger) and prevents the oil from thinning out too much at high temperatures.
- VI improvers do have a couple of negative features. The additives are large (high molecular weight) polymers, which makes them susceptible to being chopped or cut up into small pieces by machine components (shearing forces). Gears are notoriously hard on VI-improver additives.
- Permanent shearing of the VI-improver additive can cause significant viscosity losses, which can be detected with oil analysis. A second form of viscosity loss occurs due to high shearing forces in the load zone of frictional surfaces (e.g., in journal bearings).
- It is thought that the VI-improver additive loses its shape or uniform orientation and therefore loses some of its thickening ability.
- The viscosity of the oil temporarily drops within the load zone and then rebounds to its normal viscosity after it leaves the load zone. This characteristic actually aids in the reduction of fuel consumption.

There are several different types of VI improvers (olefin copolymers are common). High-quality VI improvers are less susceptible to permanent shear loss than low-cost, low-quality VI improvers.

Anti-wear (AW) Agents

These additives are typically used to protect machine parts from wear and loss of metal during boundary lubrication conditions. They are polar additives that attach to frictional metal surfaces. They react chemically with the metal surfaces when metal-to-metal contact occurs in conditions of mixed and boundary lubrication.

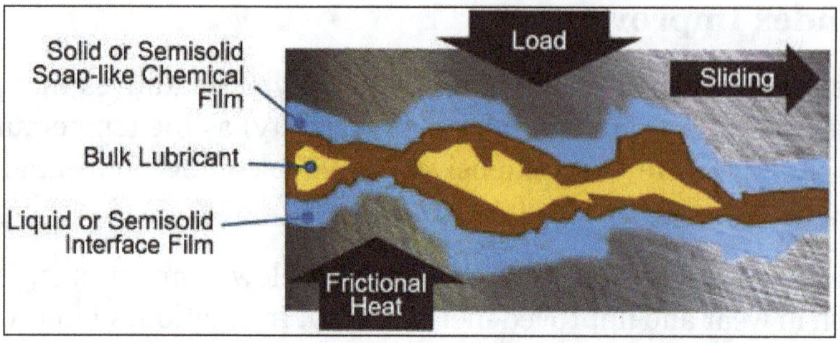

They are activated by the heat of contact to form a film that minimizes wear. They also help protect the base oil from oxidation and the metal from damage by corrosive acids. These additives become "used up" by performing their function, after which adhesive wear damage will increase. They are typically phosphorus compounds, with the most common being zinc dialkyldithiophosphate (ZDDP).

There are different versions of ZDDP — some intended for hydraulic applications and others for the higher temperatures encountered in engine oils. ZDDP also has some anti-oxidant and corrosion-inhibition properties. In addition, other types of phosphorous-based chemicals are used for anti-wear protection (e.g., TCP).

Extreme Pressure (EP) Additives

These additives are more chemically aggressive than AW additives. They react chemically with metal (iron) surfaces to form a sacrificial surface film that prevents the welding and seizure of opposing asperities caused by metal-to-metal contact (adhesive wear).

They are activated at high loads and by the high contact temperatures that are created. They are typically used in gear oils and give those oils that unique, strong sulphur smell. These additives usually contain sulphur and phosphorus compounds (and occasionally boron compounds).

They can be corrosive toward yellow metals, especially at higher temperatures, and therefore should not be used in worm gear and similar applications where copper-based metals are used. Some chlorine-based EP additives exist but are rarely used due to corrosion concerns.

Anti-wear additives and extreme pressure agents form a large group of chemical additives that carry out their function of protecting metal surfaces during boundary lubrication by forming a protective film or barrier on the wear surfaces. As long as the hydrodynamic or elastohydrodynamic oil film is maintained between the metal surfaces, boundary lubrication will not occur and these boundary lubrication additives will not be required to perform their function. When the oil film does break down and asperity contact is made under high loads or high temperatures, these boundary lubrication additives protect the wearing surfaces.

Detergents

Detergents perform two functions. They help to keep hot metal components free of deposits (clean) and neutralize acids that form in the oil. Detergents are primarily used in engine oils and are alkaline or basic in nature.

They form the basis of the reserve alkalinity of engine oils, which is referred to as the base number (BN). They are typically materials of calcium and magnesium chemistry. Barium-based detergents were used in the past but are rarely used now.

Since these metal compounds leave an ash deposit when the oil is burned, they may cause unwanted residue to form in high-temperature applications. Due to this ash concern, many OEMs are specifying low-ash oils for equipment operating at high temperatures. A detergent additive is normally used in conjunction with a dispersant additive.

Dispersants

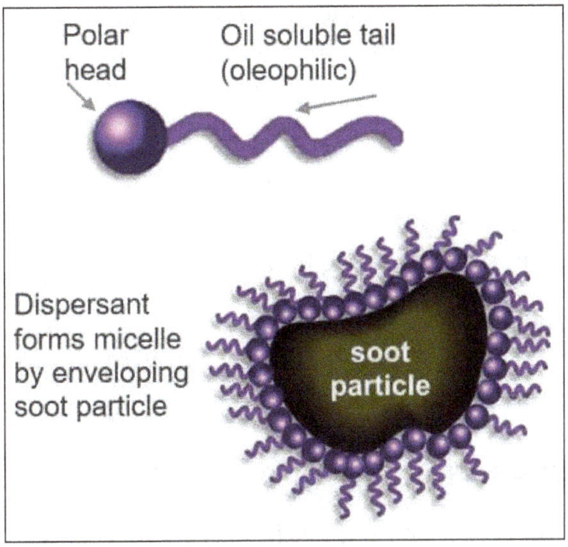

Dispersants are mainly found in engine oil with detergents to help keep engines clean and free of deposits. The main function of dispersants is to keep particles of diesel engine soot finely dispersed or suspended in the oil (less than 1 micron in size).

The objective is to keep the contaminant suspended and not allow it to agglomerate in the oil so that it will minimize damage and can be carried out of the engine during an oil change. Dispersants are generally organic and ashless. As such, they are not easily detectable with conventional oil analysis.

The combination of detergent/dispersant additives allows more acid compounds to be neutralized and more contaminant particles to stay suspended. As these additives perform their functions of neutralizing acids and suspending contaminants, they will eventually exceed their capacity, which will necessitate an oil change.

Anti-foaming Agents

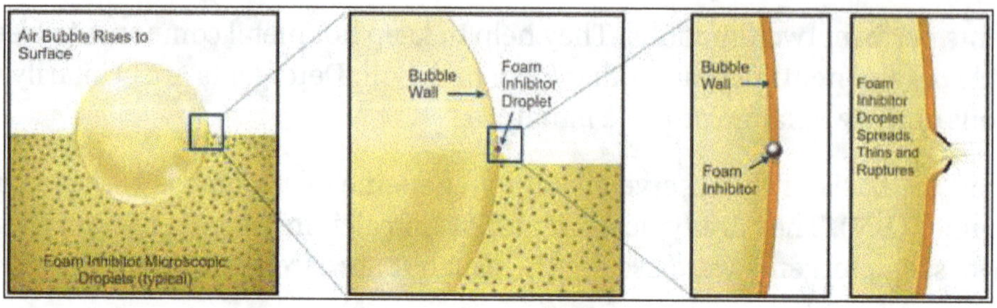

The chemicals in this additive group possess low interfacial tension, which weakens the oil bubble wall and allows the foam bubbles to burst more readily. They have an indirect effect on oxidation by reducing the amount of air-oil contact.

Some of these additives are oil-insoluble silicone materials that are not dissolved but rather dispersed finely in the lubricating oil. Very low concentrations are usually required. If too much anti-foaming additive is added, it can have a reverse effect and promote further foaming and air entrainment.

Friction Modifiers

Friction modifiers are typically used in engine oils and automatic transmission fluids to alter the friction between engine and transmission components. In engines, the emphasis is on lowering friction to improve fuel economy.

In transmissions, the focus is on improving the engagement of the clutch materials. Friction modifiers can be thought of as anti-wear additives for lower loads that are not activated by contact temperatures.

Pour Point Depressants

The pour point of an oil is approximately the lowest temperature at which an oil will remain fluid. Wax crystals that form in paraffinic mineral oils crystallize (become solid) at low temperatures. The solid crystals form a lattice network that inhibits the remaining liquid oil from flowing. The additives in this group reduce the size of the wax crystals in the oil and their interaction with each other, allowing the oil to continue to flow at low temperatures.

Demulsifiers

Demulsifier additives prevent the formation of a stable oil-water mixture or an emulsion by changing the interfacial tension of the oil so that water will coalesce and separate more readily from the oil. This is an important characteristic for lubricants exposed to steam or water so that free water can settle out and be easily drained off at a reservoir.

Emulsifiers

Emulsifiers are used in oil-water-based metal-working fluids and fire-resistant fluids to help create a stable oil-water emulsion. The emulsifier additive can be thought of as a glue binding the oil and water together, because normally they would like to separate from each other due to interfacial tension and differences in specific gravity.

Biocides

Biocides are often added to water-based lubricants to control the growth of bacteria.

Tackifiers

Tackifiers are stringy materials used in some oils and greases to prevent the lubricant from flinging off the metal surface during rotational movement.

To be acceptable to blenders and end users alike, the additives must be capable of being handled in conventional blending equipment, stable in storage, free of offensive odor and be nontoxic by normal industrial standards. Since many are highly viscous materials, they are generally sold to the oil formulator as concentrated solutions in a base oil carrier.

A couple of key points about additives:

- More additive is not always better: The old saying, "If a little bit of something is good, then more of the same is better," is not necessarily true when using oil additives.
- As more additive is blended into the oil, sometimes there isn't any more benefit gained, and at times the performance actually deteriorates. In other cases, the performance of the additive doesn't improve, but the duration of service does improve.
- Increasing the percentage of a certain additive may improve one property of an oil while at the same time degrade another. When the specified concentrations of additives become unbalanced, overall oil quality can be affected.
- Some additives compete with each other for the same space on a metal surface. If a high concentration of an anti-wear agent is added to the oil, the corrosion inhibitor may become less effective. The result may be an increase in corrosion-related problems.

How Oil Additives get Depleted

It is very important to understand that most of these additives get consumed and depleted by:

- "Decomposition" or breakdown.

- "Adsorption" onto metal, particle and water surfaces.
- "Separation" due to settling or filtration.

The adsorption and separation mechanisms involve mass transfer or physical movement of the additive. For many additives, the longer the oil remains in service, the less effective the remaining additive package is in protecting the equipment.

When the additive package weakens, viscosity increases, sludge begins to form, corrosive acids start to attack bearings and metal surfaces, and/or wear begins to increase. If oils of low quality are used, the point at which these problems begin will occur much sooner.

It is for these reasons that top-quality lubricants meeting the correct industry specifications (e.g., API engine service classifications) should always be selected. The following table can be used as a guide for a more thorough understanding of additive types and their functions in engine oil formulations.

Surface Protective Additives Engine Lubricants			
Additive type	Purpose	Typical compounds	Functions
Anti-Wear Agent	Reduce friction and wear, and prevent scoring and seizure.	Zinc dithiophosphates, organic phosphates and acid phosphates; organic sulphur and chlorine compounds; sulphurized fats, sulfides and disulfides.	Chemical reaction with the metal surface to form a film with lower shear strength than the metal, thereby preventing metal-to-metal contact.
Corrosion and Rust Inhibitor	Prevent corrosion and rusting of metal parts in contact with the lubricant.	Zinc dithiophosphates, metal phenolates, basic metal sulfonates, fatty acids and amines.	Preferential adsorption of polar constituent on metal surface to provide a protective film and/or neutralization of corrosive acids.
Detergent	Keep surfaces free of deposits and neutralize corrosive acids.	Metallo-organic compounds of barium, calcium and magnesium phenolates, phosphates and sulfonates.	Chemical reaction with sludge and varnish precursors to neutralize them and keep them soluble.
Dispersant	Keep insoluble soot dispersed in the lubricant.	Polymeric alkylthiophosphonates and alkylsuccinimides, organic complexes containing nitrogen compounds.	Contaminants are bonded by polar attraction to dispersant molecules, prevented from agglomerating and kept in suspension due to solubility of dispersant.
Friction Modifier	Alter coefficient of friction.	Organic fatty acids and amines, lard oil, high molecular weight organic phosphorus and phosphoric acid esters.	Preferential adsorption of surface-active materials.

Performance Additives Engine Lubricants			
Pour Point Depressant	Enable lubricant to flow at low temperatures.	Alkylated naphthalene and phenolic polymers, polymethacrylates.	Modify wax crystal formation to reduce interlocking.
Seal Swell Agent	Swell elastomeric seals.	Organic phosphates, aromatics, halogenated hydrocarbons.	Chemical reaction with elastomer to cause slight swell.
Viscosity Improver	Reduce the rate of viscosity change with temperature.	Polymers and copolymers of methacrylates, butadiene olefins and alkylated styrenes.	Polymers expand with increasing temperature to counteract oil thinning.
Lubricant Protective Additives Engine Lubricants			
Anti-Foaming	Prevent lubricant from forming a persistent foam.	Silicone polymers and organic copolymers.	Reduce surface tension to speed collapse of foam.
Anti-Oxidant	Retard oxidative decomposition.	Zinc dithiophosphates, hindered phenols, aromatic amines, sulphurized phenols.	Decompose peroxides and terminate free-radical reactions.
Metal Deactivator	Reduce catalytic effect of metals on oxidation rate.	Organic complexes containing nitrogen or sulphur, amines, sulphides and phosphites.	Form inactive film on metal surfaces by complexing with metallic ions.

It is evident from the information above that there is a lot of chemistry occurring in most of the oils that are used to lubricate equipment. They are complicated mixtures of chemicals that are in balance with one another and need to be respected. It is for those reasons that the mixing of different oils and adding additional lubricant additives should be avoided.

After-market Additives and Supplemental Oil Conditioners

There are hundreds of chemical additives and supplemental lubricant conditioners available. In certain specialized applications or industries, these additives may have a place in the improvement of lubrication.

However, some manufacturers of supplemental lubricants will make claims about their products that are exaggerated and/or unproven, or they fail to mention a negative side effect that the additive may cause. Take great care in the selection and application of these products, or better still, avoid using them. If you want a better oil, buy a better oil in the first place and leave the chemistry to the people who know what they are doing.

Often oil and equipment warranties are voided with the use of after-market additives because the final formulation has never been tested and approved.

When considering the use of an after-market additive to solve a problem, it is wise to remember the following rules:

- Rule 1: An inferior lubricant cannot be converted into a premium product simply by the inclusion of an additive. Purchasing a poor-quality finished oil and attempting to overcome its poor lubricating qualities with some special additive is illogical.

- Rule 2: Some laboratory tests can be tricked into providing a positive result. Some additives can trick a given test into providing a passing result. Often multiple oxidation and wear tests are run to obtain a better indication of the performance of an additive. Then actual field trials are performed.
- Rule 3: Base oils can only dissolve (carry) a certain amount of additive. As a result, the addition of a supplemental additive into an oil having a low level of solubility or being already saturated with additive may simply mean that the additive will settle out of the solution and remain in the bottom of the crankcase or sump. The additive may never carry out its claimed or intended function.

If you choose to use an after-market additive, before adding any supplemental additive or oil conditioner to a lubricated system, take the following precautions:

- Determine whether an actual lubrication problem exists. For instance, an oil contamination problem is most often related to poor maintenance or inadequate filtration and not necessarily poor lubrication or poor-quality oil.
- Choose the right supplemental additive or oil conditioner. This means taking the time to research the makeup and compatibility of the various products on the market.
- Insist that factual field-test data is made available that substantiates the claims made regarding the product's effectiveness.
- Consult a reputable, independent oil analysis laboratory. Have the existing oil analyzed at least twice before adding a supplemental additive. This will establish a reference point.
- After the addition of the special additive or conditioner, continue to have the oil analyzed on a regular basis. Only through this method of comparison can objective data regarding the effectiveness of the additive be obtained.

There is a great deal of controversy surrounding the application of supplemental additives. However, it is true that certain supplemental lubricant additives will reduce or eliminate friction in some applications such as machine tool ways, extreme pressure gear drives and certain high-pressure hydraulic system applications.

Lubricant Selection

Before the proper lubricant can be selected, the tribological system must be identified. This system includes the type of motion, speed, temperatures, load and the operating environment.

Once these system parameters are identified, the lubrication engineer (or tribo-engineer) can utilize different lubricant chemistries to select a lubricant that will optimize the performance of the application. Because each chemistry has advantages and disadvantages, it is important to choose the appropriate one to address each of the tribological system parameters.

In addition, the lubrication engineer must analyze the application based on the identified tribological system. This analysis includes elements such as speed factors, elastohydrodynamic (EHD) lubrication, bearing-life calculations, extreme-pressure lubrication, emergency lubrication and various special application requirements.

Type of Motion

The first parameter of the tribological system involves the type of motion. The motion may be sliding, which would require the hydrodynamic lubrication theory for its analysis, or rolling, in which case EHD lubrication theory would be applied.

Combined sliding and rolling is also a possible form of motion in certain rolling-element bearings including the tapered roller bearing. Sliding in the rib area can occur in this bearing, but the rolling elements roll on the raceway surfaces. Lubricant protection of these types of motion can be optimized with specific chemistries. Some lubricant chemistry is effective in sliding contacts but do not perform as well in rolling contacts.

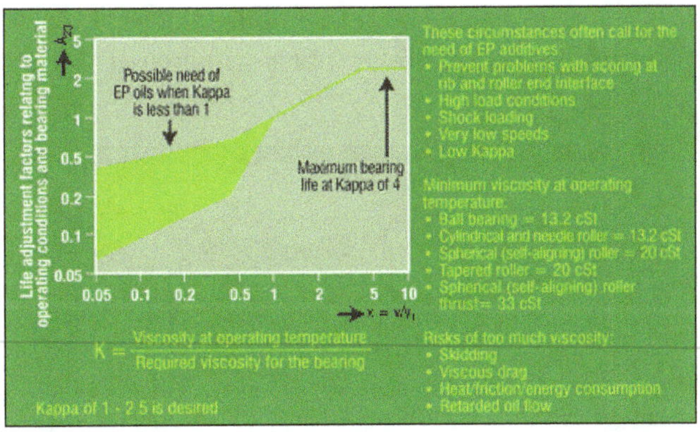

k Values.

Speed

Speed is the second parameter on the tribological system. Speed of roller element bearings can be broken into the general ranges: fast, moderate and slow. Specific ranges for each of these speed categories can be set by using the speed factor, as defined in equation:

$$\text{Speed factor} = n \cdot d_m$$

n = operating speed, rpm

$$d_m = \text{mean bearing diameter, mm}$$
$$= (\text{ID} + \text{OD}) / 2$$
$$\text{D} = \text{inside diameter, mm}$$
$$\text{OD} = \text{outside diameter, mm}^2$$

The Stribeck curve is a graph showing the relationship between coefficient of friction and the dimensionless number hn/P, where h is dynamic viscosity, n is speed, and P is the load per unit of projected area. According to this curve, there is an optimum speed for the lubricated contact. Knowing the speed of the contact, a lubricant can be selected with the optimum physical attributes to minimize friction.

Temperature

The third tribological parameter is temperature. All lubricants have specific temperature ranges for optimal performance. Many lubricants have a broad operational temperature range; however, some lubricants are more suited for lower temperatures. For example, there are some greases with synthetic hydrocarbon base oil and barium-complex thickener that can operate at temperatures as low as -60 °C.

Other lubricants are designed for high-temperature applications, such as greases with perfluorinated aliphatic ether base oil thickened with polytetrafluoroethylene (PTFE) that can lubricate an oven chain bearing at 220 °C for more than 15,000 hours. Knowing the temperature of the tribological system enables the engineer to select a lubricant that will provide optimum operating life and performance at the application temperature.

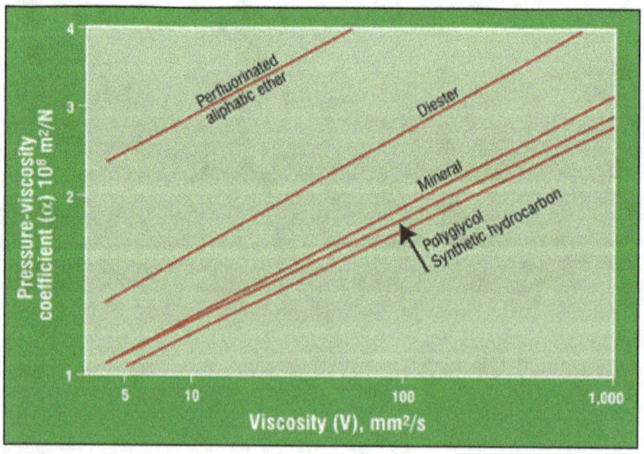

Pressure-viscosity Coefficients for Five Oils over the Pressure Range 0 to 2,000 bar.

Load

Load, the fourth parameter, is an important factor affecting the lubricant requirement. A light load may indicate the application is sensitive to frictional torque, and therefore a lubricant would have to be selected to minimize the fluid friction while still providing protection from metal-to-metal friction. On the opposite end is a heavily loaded application, which could require specific additives to help protect from pitting, galling and extreme wear.

Operating Environment

The last parameter of the tribological system is the application's operating environment. If the environment includes moisture or water, the lubricant must provide good anticorrosion properties as well as resistance to water washout or contamination. If the application operates in a vacuum or partial vacuum, the atmospheric pressure of the application must be within the operational limits of the lubricant and above its vapor pressure at the operating temperature.

If the application requires the presence of certain chemical liquids or vapors, the selected lubricant must be resistant to these chemicals. Even an ideal environment, such as a computer room or clean-room processing facility, could have specific requirements for noise-reducing lubricants in rolling-element or instrument bearings.

Tribological Analysis Theories

The five parameters of the tribological system must be taken into consideration and analyzed for the best lubricant to be selected for the application. However, the information obtained by defining the tribological system parameters also provides data for further in-depth technical analysis.

EHD Lubrication Theory

An important type of analysis involves the lubrication theory for rolling-element bearings. Elastohydrodynamic (EHD) lubrication theory, sometimes referred to as EHL, is used to identify a lubricant's film thickness in a rolling contact. A roller and a raceway can illustrate the factors affecting the minimum film thickness, ho, in the area of rolling contact. Assuming that both surfaces are perfectly smooth, we can define the minimum film thickness, ho in a rolling contact situation according to equation below which forms the basis for EHD lubrication theory:

$$h_o = \left(\left[0.1 \times \acute{a}^{0.6} \times (\eta v) 0.07\right]\right) / \left\{(1/r_1) + (1/r_2)^{0.43} (Q/1)^{0.13}\right\} \left\{E / \left[1 - (1/m^2)\right]\right\}^{0.03}$$

Where,

h_o = minimum lubricating films thickness in area of rolling contact, μm

α = pressure - viscosity, mm^2/N

η = dynamic viscosity, mPa• s

$v = (V_1 + V_2)/2$, m/s

$V1$ = roller velocity, m/s

V_2 = raceway velocity, m/s

r_1 = roller radium, mm

r_2 = inner ring raceway radius, mm

Q = roller load, N

L = roller length, mm

E = modulus of elasticity, N/mm^2

= 2.08×10^5 for steel

$1/m$ = Poisson's ration

Lubrication engineers use the EHD lubrication theory to select the proper viscosity of the lubricant. Each of the variables in equation above has a specific impact on the ultimate film thickness. Most of these variables are under the control of the designer of the application, but several may also be under the control of the lubrication engineer. One of the lubrication engineer's primary interests is how a change in a specific variable will affect the magnitude of the film thickness.

From equation above, it can be determined that if the pressure-viscosity coefficient (a) is doubled, there is an increase in the film thickness by 51 percent. Knowing the pressure-viscosity coefficient of the different lubricant chemistries, the lubrication engineer can alter the film thickness by changing the lubricant chemistry. The remaining physical characteristics of the lubricant are not changed.

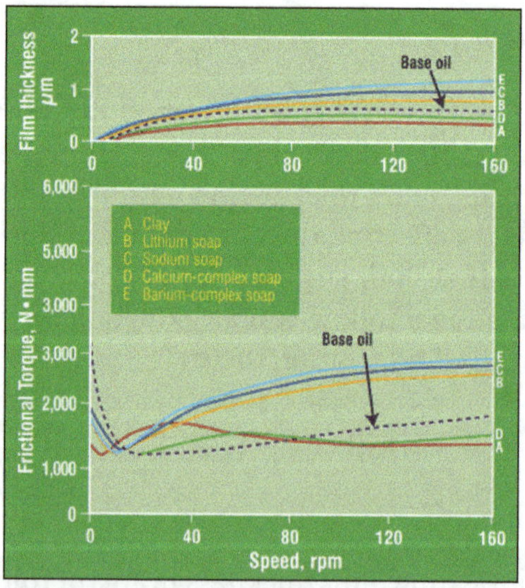

Film Thickness and Frictional Torque for Base Oil and Five Greases as Functions of Bearing Speed.

Another lubricant-related variable in equation above is the dynamic viscosity, h. The dynamic viscosity can be directly related to the kinematic viscosity, and if it is doubled, will increase film thickness by 62 percent. By doubling the velocity of the roller bearing, the film thickness of the lubricant can again be increased by 62 percent. The lubrication engineer has no control over the speed of the application, but knowing how the speed affects the film thickness is important to the selection of a lubricant when the application has variable-speed capability. Additional variables have a lesser impact on the film thickness of the lubricant.

While the information in Equation above is important to the lubricant selection criteria, the EHD film thickness is not used directly because in reality, surfaces are not perfectly smooth. Instead of minimum film thickness, the lubrication engineer works with the specific film thickness, l, defined as the ratio between the average or mean EHD film thickness, a, and the composite surface roughness of the rolling contacts s.

$$\lambda = \mathrm{ho}/\sigma$$

Where,

$$\lambda = \text{specific fims thicness}$$

$$\sigma = \text{composite roughness of the two contact surfaces}$$

$$= \left(\sigma_1^{\,2} + \sigma_2^{\,2}\right)^{1/2}$$

$$\sigma_1 = \text{rms roughness of raceway}, \mu\text{m}$$

$$\sigma_2 = \text{rms roughness of roller}, \mu\text{m}$$

From equation above, it is clear that if the specific film thickness is close to zero, there will be a dramatic increase in the metal-to-metal contact at the friction point. This increased metal contact produces unacceptable wear. When l=1, the bearing will have only partially separated the metal asperities and some metal-to-metal contact will still be occurring.

It is at this point that the transition from boundary lubrication into the mixed lubrication occurs. As the specific film thickness increases above 1 and approaches values greater than 3, there will be a decrease in metal-to-metal contact with an associated decrease in wear. When l = 3 to 4 or greater, there is full metal separation between the peaks, indicating full fluid film lubrication and an absence of wear.

Lambda values greater than 4 are possible and sometimes desirable, particularly where variable speeds and or shock loading is present. However as l increases beyond 4, the internal fluid friction may increase and generate excessive heating and energy consumption depending upon the relative bearing speed and oil viscosity.

Required Viscosity

Research sponsored by the bearing manufacturers has made it possible to calculate the minimum required viscosity of the lubricant V1 to obtain separation of the moving surfaces. The nomograph in figure can be used to determine the value of V1 from the mean bearing diameter, dm, and the operating speed, n.

For example, enter the bottom of the chart with the average diameter of the rolling-element bearing (100 mm), then proceed vertically to the speed of the bearing (2,000 rpm). Drawing a line horizontally to the left then provides the required viscosity (10mm^2/s) to obtain l = 1 in this specific application at the given operating temperature.

Knowing the required minimum viscosity, V1, at operating temperature, the lubrication engineer can then select the appropriate lubricant based on the specific viscosity, k, which is defined as the ratio of the actual viscosity, V, of the selected lubricant to the required minimum viscosity, V1. Given this information, the engineer will attempt to select a lubricant to meet the full fluid film lubrication regime for the application. If this is not possible, then the next available option is to select a lubricant that provides the best protection to the application. k relates to both lubricant selection and expected bearing life. At k values below 1, it is generally accepted that EP additives will be required to mitigate the effects of boundary lubrication conditions. As k approaches 1, the life expectancy approaches the rated life expectancy L of the bearing according to DIN ISO 281. At k values above 1, it is in fact possible to exceed the L value of the bearings, perhaps by as much as 2.5 times.

As k approaches 4, the bearing life approaches a maximum (all else being equal) while k values above 4 may cause increase fluid friction, viscous drag, ball skidding and other undesirable effects. It is generally accepted that ê values in the 1 to 2.5 range are optimal for most bearing applications.

If the application must be operated in the boundary lubrication regime, then specific additives must be provided to protect the metal contact points. If the application is operating in a lubrication regime that approaches full fluid film lubrication, these extreme-pressure additives can be eliminated from the lubricant.

Points of Consideration

The EHD theory is a valuable tool in guiding the lubrication engineer toward selecting the proper lubricant. If the lubrication engineer uses the specific viscosity as part of the selection criteria, the assumptions used in the analysis must be taken into consideration, along with the controllable variables. Two specific points should be considered:

- Calculation of film thickness (ho), specific film thickness (l), and the chart used to determine required viscosity (V1) are all based on the pressure-viscosity coefficient (a) of mineral oil. The pressure-viscosity coefficient for mineral oil and four other oils over the pressure range zero to 2,000 bar.
- The oil is considered to be the only lubricating component of the lubricant. In

a grease, it is believed that the thickener system does not contribute to the lubricating film thickness. However, the thickener system can have a significant effect on film thickness.

Figure documents research identifying the effect of a grease thickener system on the actual film thickness. The upper chart shows that a clay-thickened grease will reduce the film thickness of the application to almost 50 percent of the film thickness provided by the base oil alone. A grease thickened with a barium-complex soap has the opposite effect of almost doubling the film thickness relative to the base oil alone.

However, as shown in the lower chart in figure, the advantage of increased load-carrying film thickness comes with the disadvantage of frictional torque. While a clay-thickened grease reduces the film thickness, it does have the benefit of reducing frictional torque in the bearing. The opposite effect is observed for grease thickened with a barium-complex soap.

Speed Factor and Bearing Design

Speed is another parameter of the tribological system that requires further detailed analysis. The calculated speed factor for a rolling-element bearing, which is defined as the product of bearing speed (n) and mean bearing diameter (dm), is not consistent for all bearings with the same dimensions and speed. The lubrication engineer must apply a correction to the calculated speed factor, depending on the bearing design.

Lubricant Selection Criteria

Tribological System

Analysis of the tribological system for a given application is essential to the selection of the appropriate lubricant.

- Service temperature range,
- Speed factor (ndm),
- Hydrodynamic lubrication,
- Elastohydrodynamic lubrication,
- Extreme pressure,
- Emergency lubrication,
- Fretting.

Many applications have special requirements that go beyond the tribological system that must be taken into consideration. Some applications are limited to oils, while others require a grease. Applications that involve the use of sintered bearings or special sealing arrangements will require additional analysis. Material compatibility is another important issue.

Other Requirements

Lubricant selection can also be affected by a variety of other specialized requirements:

- Design life.
- Lubrication equipment.
- Acceptable relubrication intervals.
- Cost.
- Special certifications such as NSF registration.
- Biodegradability.

Multipurpose lubricants cannot provide satisfactory service in current demanding environments.

Lubricant performance must be optimized to meet the increasing demands of modern industry.

The first step in selecting the best lubricant for a given application is to define the tribological system. With a fully defined tribological system in place, the next step is theoretical analysis. Selection of a lubricant based on EHD lubrication analysis or analysis of any other discrete parameter is inappropriate, because such analyses focus only on a subset of the tribological system.

Hydrostatic Lubrication

Hydrostatic lubrication is essentially a form of hydrodynamic lubrication in which the metal surfaces are separated by a complete film of oil, but instead of being self-generated, the separating pressure is supplied by an external oil pump. Hydrostatic lubrication depends on the inlet pressure of lube oil and clearance between the metal surfaces, whereas in hydrodynamic lubrication it depends on the relative speed between the surfaces, oil viscosity, load on the surfaces, and clearance between the moving surfaces.

Example: The cross head pin bearing or gudgeon pin bearing in two stroke engines employs this hydrostatic lubrication mechanism. In the cross head bearing, the load is very high and the motion is not continuous as the bearing oscillation is fairly short.

Thus hydrodynamic lubrication cannot be achieved. Under such conditions, hydrostatic lubrication offers the advantage. The oil is supplied under pressure at the bottom of bearing. The lube oil pump pressure is related to the load, bearing clearance, and thickness of the oil film required, but is usually in the order of 35-140 kg/cm².

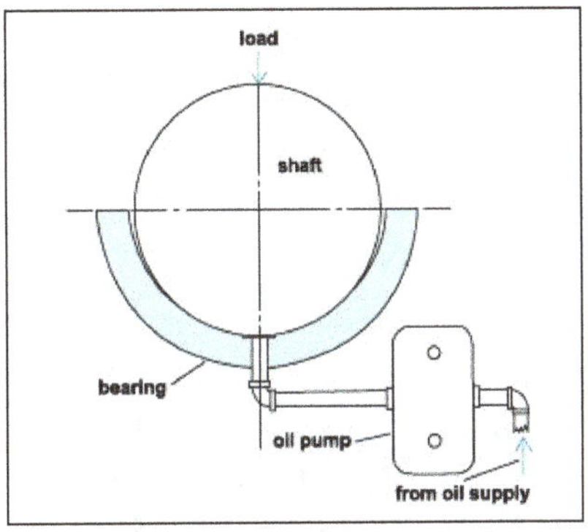

Hydrostatic Lubrication.

ElastoHydrodynamic Lubrication

Elastohydrodynamic Lubrication – or EHL – is a lubrication regime (a type of hydrodynamic lubrication (HL)) in which significant elastic deformation of the surfaces takes place and it considerably alters the shape and thickness of the lubricant film in the contact. The term underlies the importance of the elastic deflection of the bodies in contact in the development of the total lubricant film. EHL, the same way as HL, is used to decrease friction and wear in tribological contacts. It is achieved by the development of a thin lubricant film between rubbing surfaces, which separates them and decreases friction. EHL has characteristic features, such as constant film thickness and almost Hertzian contact pressure profile within the Hertzian contact area. These features have been extensively used in construction of approximate solutions of EHL theory.

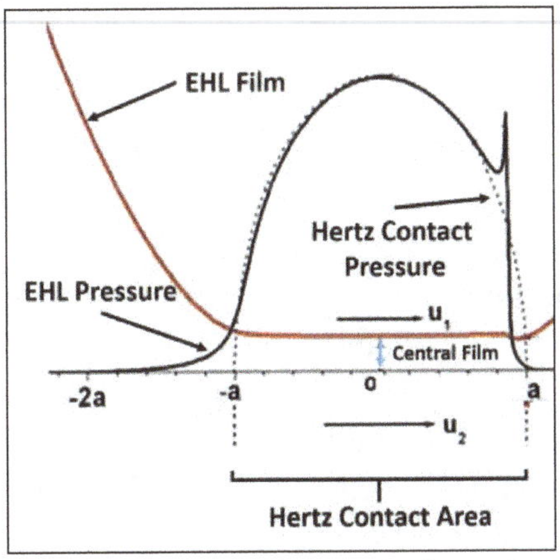

Hertz Contact Pressure vs. Elastohydrodynamic Pressure.

System of EHL Equations

A classical EHL system of equations consists of the system of Reynolds equation, film thickness and load balance equations:

$$\frac{\partial}{\partial x}\left(\frac{h^3}{\mu}\frac{\partial p}{\partial x}\right)+\frac{\partial}{\partial z}\left(\frac{h^3}{\mu}\frac{\partial p}{\partial z}\right)=6(U_1+U_2)\frac{\partial h}{\partial x}+12\frac{\partial h}{\partial t},$$

$$h=h_o+h_g+h_e+h_m,$$

$$F_N=\int_{-\infty}^{\infty} p dA$$

where h, p, μ are hydrodynamic film thickness, pressure, viscosity, and U_1 and U_2 represent the velocity of the bearing surfaces. Variables h_o, h_g, h_e, h_m represent the approach, macroscale geometry, elastic distortion of the surfaces and microscale geometry (surface roughness) correspondingly. This system of equations can be solved assuming appropriate boundary conditions to obtain unknown hydrodynamic pressure and film thickness in the contact. Typically, parameter is unknown (although sometimes it can be specified), therefore the last integral equation is needed to get the closed system of equations. F_N is the normal load applied to the contact.

The system of equations shown above can be solved analytically in certain cases, however, in general it has to be solved using numerical methods. The problem in solving the Reynolds equation comes from the film thickness equation, when the elastic deflection of the surfaces is not negligible. For a 2-D case, this term can be calculated from the following equation:

$$h_e(x,y)=\frac{2\pi}{E'}\iint\frac{p(x',y')}{\sqrt{(x-x')^2+(y-y')^2}}dx'dy'$$

where E' is the reduced elastic modulus. This equation is the analytical solution of the theory of elasticity equations for a semi-infinite body subjected to normal pressure. The most robust and fast way (in terms of iterations at least) to solve EHL system is to use a fully coupled approach and Newton's scheme. However, since the elastic distortion equation is given in the integral form, the Jacobian of the system is full which increases the demands in memory enormously. In addition, solution of the equations with full Jacobian is computationally significantly more intense compared to diagonally banded cases. Therefore, researchers worked hard to develop alternative solution methods. The two most common methods for solving EHL systems numerically are the Multi-level-Multigrid and Differential Deflection techniques. The former uses multiple grids and specific integration of the film thickness equation to build an iterative solver. The latter solves a fully coupled system of equations however, instead of using the original integral form of the film thickness equation, it considered the 2-nd derivative of it. It turns out that the use of the derivative equation allows to construct a banded Jacobian and improve the efficiency of coupled approach significantly.

Recently, a so called full system approach was proposed. In this case, a Finite Element Methods are used to calculate both Reynolds and elasticity equations in a coupled manner. This method is computationally more demanding since the subsurface volume has to be discretized to calculate elastic distortions (the fully coupled approach based on differential deflection is faster than the FEM based full system technique). Nevertheless, the approach has the advantage of flexibility since it can be developed using commercial software such as COMSOL.

Film Thickness Measurements

Since the film thickness controls the separation of the rubbing surfaces and consequently friction, researchers developed several ways to measure the hydrodynamic film in the contact. One of the most frequently used techniques is based on optical interferometry. The instrument measures the lubricant film thickness in the contact formed between a steel ball and a rotating glass disc covered by a specific layer. The lubricant film thickness at any point in the image can be accurately calculated by measuring the wavelength of light at that point. Film thicknesses down to 1 nm can be measured by this approach.

References

- Lubrication-and-lubricants, tribology-fundamentals-and-advancements: intechopen.com, Retrieved 23 April, 2020
- Lubrication-applications: techtransfer.com, Retrieved 19 August, 2020
- Basic-modes-of-lubrication: enggarena.net, Retrieved 16 May, 2020
- Types-of-lubricants: 3dinsider.com, Retrieved 28 April, 2020
- Hydrodynamic-lubrication: tribonet.org, Retrieved 25 June, 2020
- Oil-lubricant-additives-31107: machinerylubrication.com, Retrieved 05 August, 2020
- Hydrostatic-lubrication, manufacturing-technology-73568: brighthubengineering.com, Retrieved 16 June, 2020

Bearings and Its Types

4

- Bearing Modulus
- Bearing Pressure
- Babbitt
- Bridge Bearing
- Bearing Materials
- Journal Bearings
- Air Bearing
- Linear-motion Bearing
- Spherical Bearing
- Cannon Bearing
- Flexure Bearing
- Jewel Bearing
- Pressure Distribution
- Thrust Bearing
- Rolling Elements

A bearing is an element of machine that resists its motion and helps in reducing the friction between moving parts. Hydrodynamic thrust bearings use hydrodynamic lubrication for positioning and movement of axis load from shift to stationary part. All the aspects related to bearings and hydrodynamic thrust bearings have been carefully analyzed in this chapter.

In a bearing, lubrication forms a thin oil film on both the rolling surface and sliding surface to prevent metalto-metal contact. The benefits of lubricating a rolling bearing can be summarized as follows:

- Alleviation of friction and wear.
- Removal of friction heat.
- Longer bearing life.
- Rust prevention.
- Protection against contamination by foreign matter.

Table: Oil volume, friction loss and bearing temperature.

Zone	Features	Typical lubrication method
A	With an extremely low oil volume, partial metal-to-metal contact occurs between the rolling elements and raceway surface, possibly leading to abnormal wear and bearing seizure.	--
B	A uniform, non-interrupted oil film is formed. Friction is minimal and bearing temperature is kept low.	Grease lubrication Oil mist lubrication Air-oil lubrication
C	Even with a greater oil volume, heat generation and cooling are in balance.	Circulating lubrication
D	Temperature increase is constant regardless of oil volume.	Circulating lubrication
E	A further increase in oil volume contributes to a significant cooling effect, and the bearing temperature drops.	Forced circulating lubrication Jet lubrication

To achieve the full lubricating effect, it is necessary to use the lubricating system best suited for design operating conditions; select a quality lubricant; remove dust from the lubricant; and design an appropriate sealing structure to prevent contamination by foreign matter as well as lubricant leakage.

The main spindle of a machine tool usually uses an extremely low volume of lubricant so heat generation from stirring of lubricant is minimal. Figure summarizes the interrelation among oil volume in a bearing, friction loss, and bearing temperature.

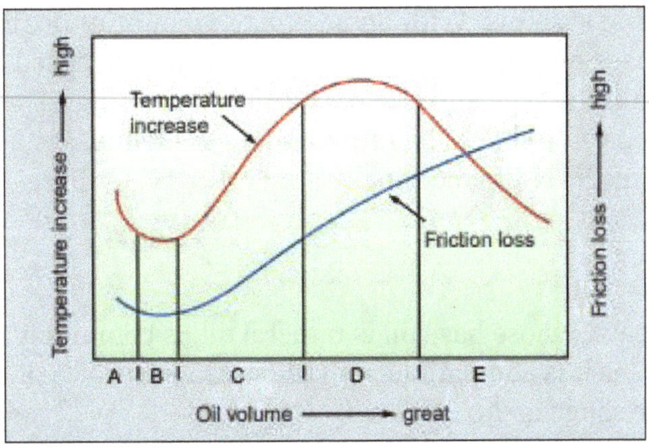

The lubrication methods available for bearings on a machine tool include grease lubrication, oil mist lubrication, air-oil lubrication, and jet lubrication. Each method has unique advantages. Therefore, a lubricating system should be selected that best suits the lubrication requirements. Tables summarize the features of various lubrication methods.

Table: Evaluation of various lubricating systems and criteria.

Lubrication method Criterion	Grease lubrication	Oil mist lubrication	Air-oil lubrication	Jet lubrication
Handling	Excellent	Good	Good	Fair
Reliability	Good	Fair	Good	Excellent
Temperature increase	Fair	Fair	Good	Excellent
Cooling effect	Poor	Fair	Good	Excellent
Sealing structure	Fair	Good	Good	Poor
Power loss	Good	Good	Good	Poor
Environmental contamination	Good	Poor	Fair	Good
Allowable dmn value	140×104	220 ×104	250 ×104	400 ×104

Grease Lubrication

Grease lubrication is the most common, as it simplifies the main spindle structure more than other lubricating systems. With an adequate amount of quality grease prefilled, this system can be used over a wide. The allowable maximum speed of a greaselubricated bearing varies with the type and size of bearing: With a high-speed angular contact ball bearing, the dmn value should be 140×10^4 as a guideline. For applications exceeding this range, consult NTN Engineering.

Grease Types

A lithium-based grease, whose base oil is mineral oil, is commonly used as a lubricant for rolling bearings, as it is compatible with the working surface of the bearing. Its operating temperature range is -30 °C to 130 °C.

When the temperature increase must be limited, as with the main spindle of a machine tool, NTN recommends the use of a synthetic-oil-based grease (diester, diester+mineral oil) whose consistency is NLGI grade 1 or 2.

Table: Lists technical data for greases commonly used for the main spindle of a machine tool.

Grease brand	ISOFLEX LDS18	Multemp PS2	Multemp LRL3	ISOFLEX NBU15	Stabllugs NBU 8EP	MP-1
Thickener	Li soap			Ba complex soap		Urea
Base oil	Synthetic oil	Diester+mineral oil	Synthetic oil	Ester	Mineral oil	Synthetic oil
Base oil viscosity (40 °C) mm²/S	16	15.3	37.3	20	105	40.6
Dropping point °C	>180	190	208	>200	220	254
Operating temperature range °C	-60~+130	−55~+130	−40~+150	−60~+130	−35~+150	−40~+150
Application	Suitable for ball bearing	Excellent low temperature and friction characteristics	Wider operating temperature range	Most commonly used for main spindle	Suitable for roller bearings subject to large loads	Applied to ULTAGE Series grease-lubricated sealed angular contact ball bearings

Amount of Grease Required

Usually, a bearing for the main spindle of a machine tool requires a small amount of grease so heat generated by the stirring of the grease during high-speed operation is minimal. As a guideline, the amount of grease used for a main spindle bearing is given below, by type and speed of bearing.

- Angular contact ball bearing:
  - (dmn value $\leqq 65 \times 10^4$): 15% of bearing space.
  - (dmn value $> 65 \times 10^4$): 12% of bearing space.
- Cylindrical roller bearing: 10% of bearing space.
- Tapered roller bearing: 15% of bearing space.

The spaces in the bearings typically used for the main spindle are listed in dimension tables. Determine a fill amount by referring to the relevant dimension table. Before filling a bearing with grease, remove the rustproof coating from the bearing with clean wash oil and allow the bearing to dry fully. Then, fill and uniformly distribute an appropriate amount of grease in the bearing with an injector or a plastic bag, etc.

Air-oil Lubrication

Air-oil lubrication (also known as oil-air lubrication or oil and air lubrication) is widely adopted for main spindle bearings in order to cope with the higher speed and precision of the main spindles of machine tools and by ensuring more reliable lubrication.

Air-oil lubrication employs a method by which compressed air is used to feed lubricating oil in precisely controlled amounts. Generally, an air-oil lubrication unit is adopted that operates a volumetric piston-type distributor that accurately meters the required minimum amount of lubricating oil and feeds it at optimal intervals controlled by a timer.

Features of Air-oil Lubrication

Air-oil lubrication has the following advantages over conventional oil mist lubrication:

- Accurately supplies a minimal amount of oil.
- Can be adjusted to provide the proper amount of lubricant for individual bearings.
- No limitations exist regarding lubricant viscosity and extreme pressure additives.
- Compressed air helps cool the bearing.
- Variations in the distance and height of lubrication points does not affect lubrication efficiency.
- Health hazards of oil mist are minimized.
- Low consumption of lubricating oil.
- Compressed air can prevent contamination of bearing by coolant.
- The recommended oil viscosity is 10 to 32 mm2/s.

Example of an Air-oil Lubrication Unit

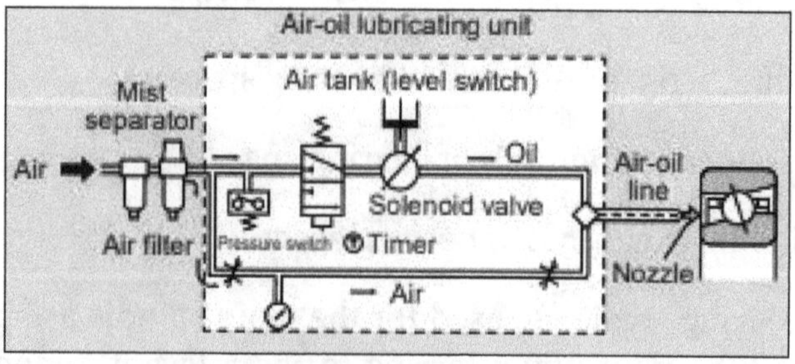

Air-oil lubricating system.

Figure shows the structure of an air-oil lubrication unit. About air-oil lubrication nozzle spacer Air-oil lubrication requires a specialized nozzle because it feeds an optimal amount of lubricating oil to the inside of the bearing by means of compressed air. A nozzle with a hole diameter of 1.0 to 1.5 mm and a length 4 to 6 times the hole diameter is recommended.

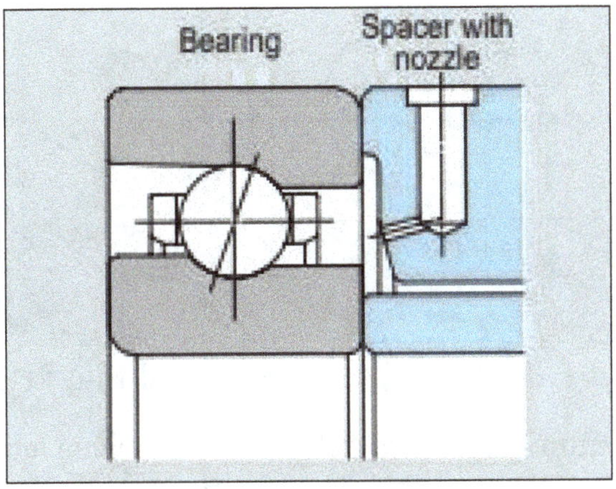

Feed system for air-oil lubrication.

Exhausting Method for Air-oil Lubrication

Air-oil lubrication is a lubricating method in which a large amount of air is used to feed lubricating oil to the inside of the bearing. Therefore, it is essential that the air fed into the bearing be allowed to escape. If the air is not smoothly exhausted, the lubricating oil will remain in the bearing and possibly contribute to bearing seizure. In the design stage, remember to allow ample space on the exhaust side of the bearing in order to increase exhaust efficiency and provide a larger oil drain hole to ensure smooth airflow. In addition, for types that allow for repositioning of the spindle, it is recommended that the shoulder dimensions of all parts be designed to prevent lubricating oil from flowing back into the bearing because of the change in attitude of the main spindle. Unnecessary dimensional differences can contribute to stagnancy of the lubricating oil.

Jet Lubrication

With this lubricating system, a high-speed jet of lubricant is injected into the bearing from the side. This is the most reliable lubricating technique and is typically used on the main spindle bearings of jet engines and gas turbines. It is currently capable of a $d_m n$ value of up to approximately 400×10^4.

When used as a lubricating system for the main spindle of a machine tool, it can minimize the temperature increase of the bearing. However, the resultant torque loss is great, as a large amount of oil is supplied to each bearing. Therefore, this arrangement requires a powerful motor to drive the main spindle. Low viscosity oil (2-3 mm^2/s) is used.

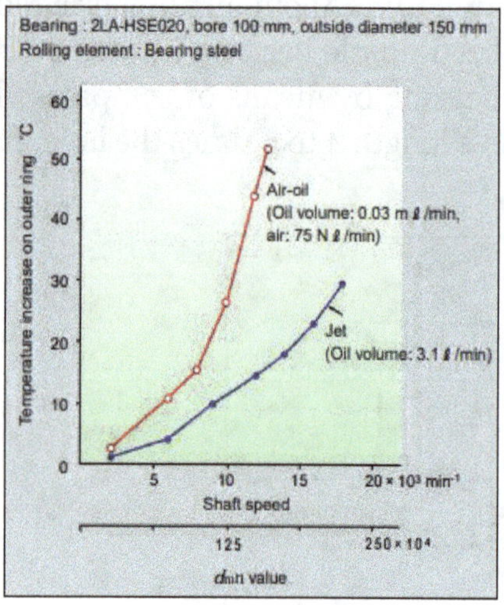

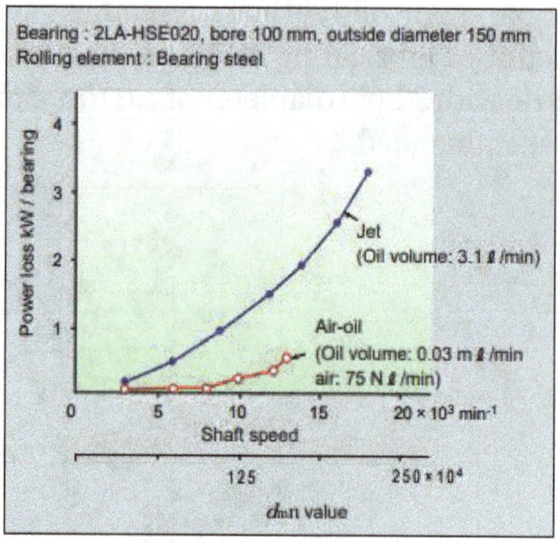

Comparison of power loss with air-oil lubrication and with jet lubrication.

Comparison of temperature increase of outer ring with air-oil lubrication and jet lubrication (The temperature increase with air-oil lubrication is relative to room temperature; the temperature with jet lubrication is relative to lubricant temperature).

Figure shows examples of the temperature increase of the same bearing with air-oil lubrication and jet lubrication, while figure graphically plots test results of power loss with the same bearing.

Bearing Modulus

Bearing modulus is a modulus used in journal bearing design. It is a dimensionless number.

Formula

Bearing Modulus (C) is C = (Zn/p) where,

- Z = oil viscosity.
- n = speed of rotation (rpm).
- p = bearing pressure(N/MM2).

For any given bearing, there is a value for indicated by C, for which the coefficient of friction is at a minimum. The bearing should not be operated at this value of bearing modulus, since a slight decrease in speed or a slight increase in pressure will make the part of a shaft or axle that rests on bearings operate in partial lubrication state resulting in high friction, heating and wear.

To prevent this, the average value of Bearing modulus should be,

$Zn/p >= 3C$

or for large fluctuations and heavy impact loads,

$Zn/p = 15C$ (approx).

Bearing Pressure

Bearing pressure is a particular case of contact mechanics often occurring in cases where a convex surface (male cylinder or sphere) contacts a concave surface (female cylinder or sphere: bore or hemispherical cup). Excessive contact pressure can lead to a typical bearing failure such as a plastic deformation similar to peening. This problem is also referred to as bearing resistance.

Hypotheses

A contact between a male part (convex) and a female part (concave) is considered when the radii of curvature are close to one another. There is no tightening and the joint slides with no friction therefore, the contact forces are normal to the tangent of the contact surface.

Moreover, bearing pressure is restricted to the case where the charge can be described by a radial force pointing towards the center of the joint.

Case of a Cylinder-cylinder Contact

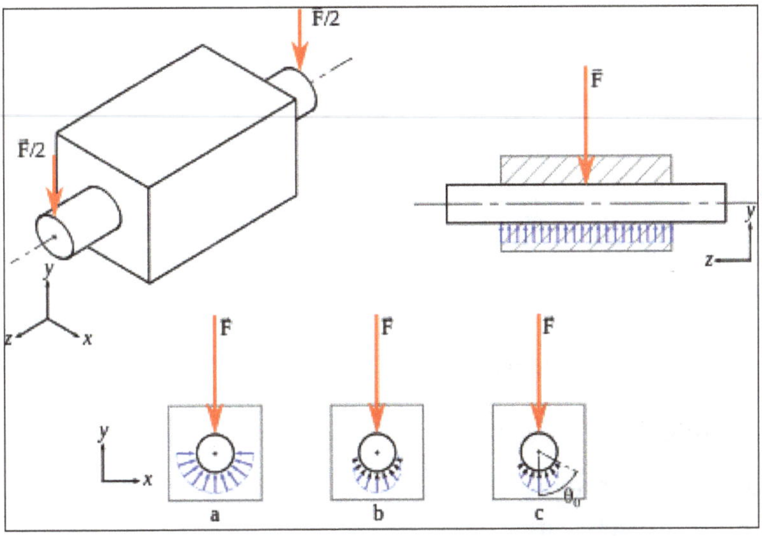

Bearing pressure for a cylinder-cylinder contact.

In the case of a revolute joint or of a hinge joint, there is a contact between a male cylinder and a female cylinder. The complexity depends on the situation, and three cases are distinguished:

- The clearance is negligible:
  - The parts are rigid bodies.
  - The parts are elastic bodies.
- The clearance cannot be ignored and the parts are elastic bodies.

By "negligible clearance", H7/g6 fit is typically meant.

The axes of the cylinders are along the z-axis, and two external forces apply to the male cylinder:

- A force $\vec{F}$ along the y-axis, the load.
- The action of the bore (contact pressure).

The main concern is the contact pressure with the bore, which is uniformly distributed along the z-axis.

Notation:

- D is the nominal diameter of both male and female cylinders.
- L the guiding length.

Negligible Clearance and Rigid Bodies

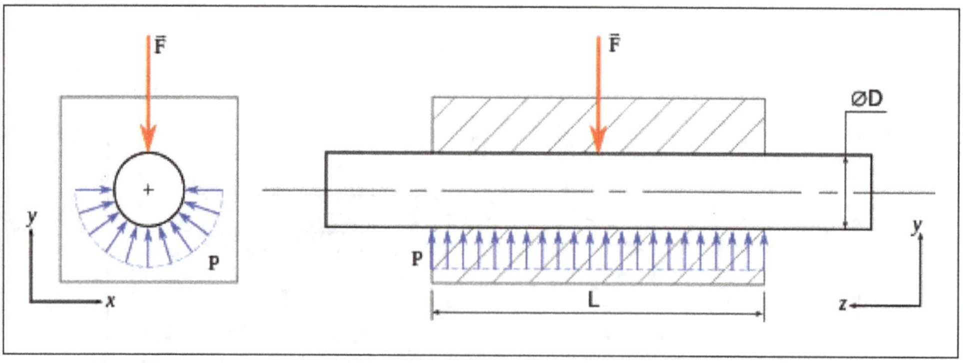

Uniform bearing pressure: case of rigid bodies when the clearing can be neglected.

In this first modeling, the pressure is uniform. It is equal to:

$$P = \frac{F}{D \times L} = \frac{\text{radial load}}{\text{projected area}}.$$

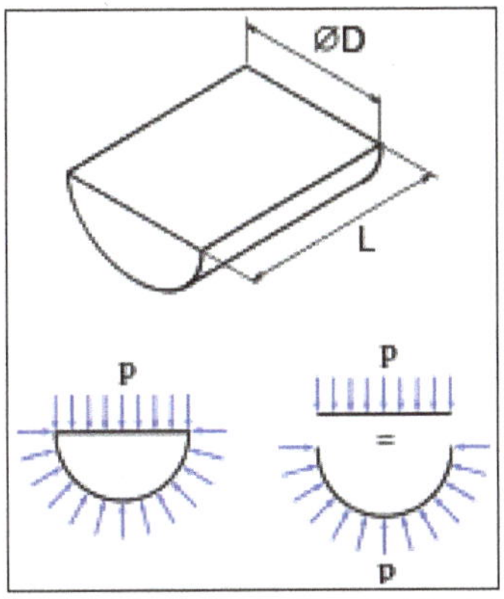

Hemicylindrical body at the equilibrium in a fluid with hydrostatic pressure.

There are two ways to obtain this result: First, we can consider a hemicylinder in a fluid, with a uniform hydrostatic pressure. The equilibrium is achieved when the resulting force on the flat surface is equal to the resulting force on the curved one. The flat surface is a D × L rectangle, therefore,

$$F = P \times (D \times L)$$

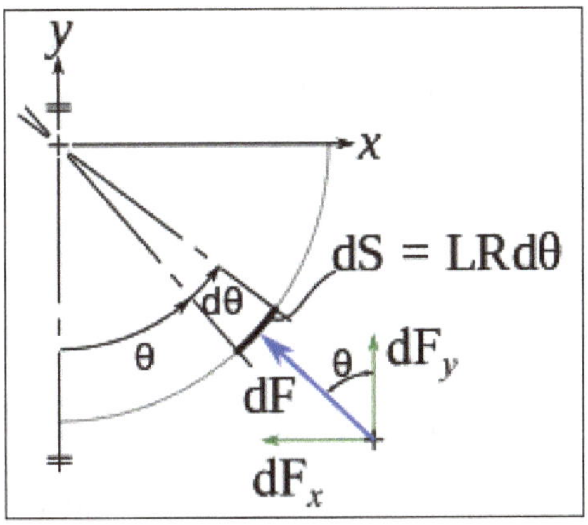

The elementary force dF, due to the pressure on a surface element dS, has two components: dF_x and dF_y.

Second, we can integrate the pressure elementary forces. Consider a small surface dS on the cylindrical part, parallel to a generating line; its length is L, and it is bound by the angles θ and θ + dθ. This small surface element can be considered as a flat rectangle which dimensions are L × (dθ × D/2). The pressure force on the surface is equal to,

$$dF = P \times dS = \tfrac{1}{2} \times P \times D \times L \times d\theta$$

The (y, z) plane is a plane of reflection symmetry, so the x compound of this force is annihilated by the force on the symmetrical surface element. The y compound of this force is equal to:

$$dF_y = \cos(\theta)\, dF = \tfrac{1}{2} \times \cos(\theta) \times P \times D \times L \times d\theta.$$

The resulting force is equal to,

$$F_y = \int_{-\pi/2}^{\pi/2} \frac{1}{2} \times P \times D \times L \times \cos(\theta) \times d\theta = \frac{1}{2} \times P \times D \times L \times [\sin(\theta)]_{-\pi/2}^{\pi/2} = P \times D \times L$$

Negligible Clearance and Elastic Bodies

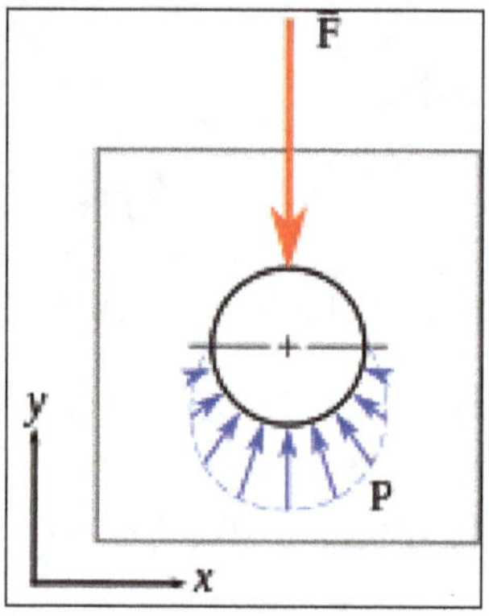

Bearing pressure with a sinusoid repartition: case of elastic bodies when the clearing can be neglected.

If it is considered that the parts deform elastically, then the contact pressure is no longer uniform and transforms to a sinusoidal repartition:

$$P(\theta) = P_{max} \cdot \cos\theta$$

with,

$$P_{max} = \frac{4}{\pi} \cdot \frac{F}{LD}.$$

This is a particular case of the following section ($\theta_o = \pi/2$).

The maximum pressure is $4/\pi \approx 1.27$ times bigger than the case of uniform pressure.

Clearance and Elastic Bodies

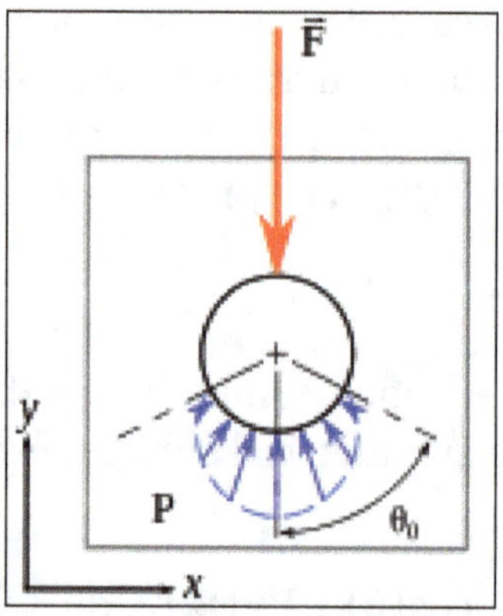

Bearing pressure in case of elastic bodies when the clearance must be taken into account.

In cases where the clearance cannot be neglected, the contact between the male part is no longer the whole half-cylinder surface but is limited to a $2\theta_0$ angle. The pressure follows Hooke's law:

$$P(\theta) = K \cdot \delta^{\alpha}(\theta)$$

where,

- K is a positive real number that represents the rigidity of the materials.
- $\delta(\theta)$ is the radial displacement of the contact point at the angle θ.
- α is a coefficient that represents the behaviour of the material:
  - $\alpha = 1$ for metals (purely elastic behaviour).
  - $\alpha > 1$ for polymers (viscoelastic or viscoplastic behaviour).

The pressure varies as:

$$A \cdot \cos\theta - B$$

where A and B are positive real number. The maximum pressure is:

$$P_{max} = \frac{4F}{LD} \times \frac{1 - \cos\theta_0}{2\theta_0 - \sin 2\theta_0}$$

the angle θ_0 is in radians.

The rigidity coefficient K and the half contact angle θ_0 cannot be derived from the theory. They must be measured. For a given system given diameters and materials, thus for given K and clearance j values, it is possible to obtain a curve $\theta_0 = f(F/(DL))$.

Relationship Between Pressure, Clearance and Contact Angle

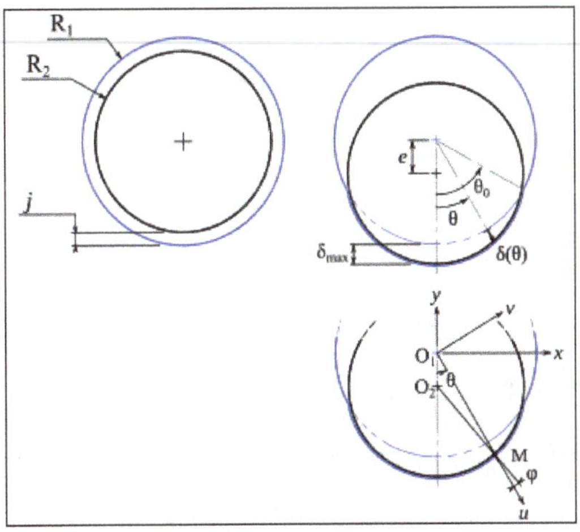

Elastic deformation in case of a male-female cylinders contact.

The part no. 1 is the containing cylinder (female, concave), the part no. 2 is the contained cylinder (male, convex); the center of the cylinder i is O_i, and its radius is R_i.

The reference position is an ideal situation where both cylinders are concentric. The clearing, expressed as a radius (not diameter), is:

$$j = R_1 - R_2.$$

Under the load, the part 2 gets in contact with the part 1, the he surfaces deform. we suppose that the cylinder 2 is rigid (no deformation), and that the cylinder 1 is an elastic body. The indentation of 2 into 1 has a depth of δ_{max}; the cylinder movement is e (excentration):

$$e = O_1O_2 = j + \delta_{max}.$$

We consider the frame at the center of the cylinder 1 (O_1, x, y). Let M be a point on the contact surface; θ is the angle (-y, O_1M). The displacement of the surface, δ, is:

$$\delta(\theta) = O_1M - R_1.$$

with $\delta(0) = \delta_{max}$. The coordinates of M are:

$$M((R_1 + \delta(\theta) \cdot \sin\theta) ; -(R_1 + \delta(\theta)) \cdot \cos\theta)$$

and the coordinates of O_2:

$$O_2(0 ; -e).$$

Consider the frame (O_1, u, v), where the axis u is (O_1M). In this frame, the coordinates are:

$$M(R_1 + \delta(\theta) ; 0)$$
$$O_2(e \cdot \cos\theta ; -e \cdot \sin\theta)$$

We know that,

$$\overrightarrow{O_1M} = \overrightarrow{O_1O_2} + \overrightarrow{O_2M}$$

thus,

$$\begin{cases} R_1 + \delta(\theta) = e \cdot \cos\theta + R_2 \cos\varphi \\ 0 = -e \cdot \sin\theta + R_2 \sin\varphi \end{cases}$$

then we use the expression of e and $R_1 = j + R_2$:

$$\begin{cases} j + R_2 + \delta(\theta) = (j + \delta_{max}) \cdot \cos\theta + R_2 \cos\varphi \\ 0 = -(j + \delta_{max}) \cdot \sin\theta + R_2 \sin\varphi \end{cases}$$

The deformations are small, as we are in the elastic domain. Thus, $\delta_{max} \ll R_1$ and therefore $|\varphi| \ll 1$, i.e.

$$\cos\varphi \simeq 1$$
$$\sin\varphi \simeq \varphi \text{ (in radians)}$$

thus

$$\begin{cases}(j+\delta_{max})\cdot\cos\theta + R_2 - (j + R_2 + \delta(\theta)) = 0\\ (j+\delta_{max})\cdot\sin\theta - R_2\cdot\varphi = 0\end{cases}$$

and

$$\begin{cases}(j+\delta_{max})\cdot\cos\theta - j - \delta(\theta) = 0\\ (j+\delta_{max})\cdot\sin\theta - R_2\cdot\varphi = 0\end{cases}$$

At $\theta = \theta_0$, $\delta(0) = 0$ and the first equation is,

$$(j+\delta_{max})\cdot\cos\theta_0 - j = 0 \Rightarrow \delta_{max} = \frac{j(1-\cos\theta_0)}{\cos\theta_0} \Leftrightarrow \cos\theta_0 = \frac{j}{j+\delta_{max}}$$

and thus

$$\left(j + \frac{j(1-\cos\theta_0)}{\cos\theta_0}\right)\cdot\cos\theta - j - \delta(\theta) = 0$$

$$\Rightarrow \delta(\theta) = j\left(\frac{\cos\theta}{\cos\theta_0} - 1\right)$$

If we use the law of elasticity for a metal ($\alpha = 1$):

$$\begin{cases}P(\theta) = K\cdot j\left(\dfrac{\cos\theta}{\cos\theta_0} - 1\right)\\ P_{max} = K\cdot j\cdot\dfrac{1-\cos\theta_0}{\cos\theta_0}\end{cases}$$

The pressure is an affine function of $\cos\theta$:

$$P(\theta) = A\cdot\cos\theta - B$$

with $A = K\cdot j/\cos\theta_0$ and $B = A\cdot\cos\theta_0$.

Case where the Clearance can be Neglected

If $j \simeq 0$ ($R_1 \simeq R_2$), then the contact is on the whole half-perimeter: $2\theta_0 \simeq \pi$ and $\cos\theta_0 \simeq 0$. The value of $1/\cos\theta_0$ rise towards infinity, thus

$$\delta(\theta) \simeq j\frac{\cos\theta}{\cos\theta_0}$$

As j and $\cos\theta_0$ both tend towards 0, the ratio $j/\cos\theta_0$ is not defined when j goes to 0. In mechanical engineering, j = 0 is an uncertain fit, it is a nonsense, both mathematically and mechanically. We are looking for a limit function,

$$P_{\pi/2}(\theta) = \lim_{\theta_0 \to \pi/2} P_{\theta_0}(\theta).$$

So, the pressure is a sinusoid function of θ:

$$P(\theta) = K \cdot \frac{j}{\cos\theta_0} \cdot \cos\theta$$

thus,

$$P(\theta) = P_{max} \cdot \cos\theta$$

with,

$$P_{max} = K \cdot \frac{j}{\cos\theta_0}.$$

Consider an infinitesimal element of surface dS bound by θ and θ + dθ. As in the case of the uniform pressure, we have

$$\begin{aligned} dF_y(\theta) &= \cos(\theta)dF = ½ \times \cos(\theta) \times P(\theta) \times D \times L \times d\theta \\ &= ½ \times \cos^2(\theta) \times P_{max} \times D \times L \times d\theta. \end{aligned}$$

When we integrate between $-\pi/2$ and $\pi/2$, the result is:

$$F = \int_{-\pi/2}^{\pi/2} F_y(\theta)d\theta = \frac{1}{2}P_{max}DL\int_{-\pi/2}^{\pi/2}\cos^2\theta d\theta$$

We know that (e.g. using the Euler's formula):

$$\int_{-\pi/2}^{\pi/2}\cos^2\theta d\theta = \frac{1}{4}[2\theta + \sin 2\theta]_{-\pi/2}^{\pi/2} = \frac{1}{2}[\theta + \sin\theta\cos\theta]_{-\pi/2}^{\pi/2} = \frac{\pi}{2}$$

therefore,

$$F = \frac{\pi}{4}P_{max}DL$$

and thus

$$P_{max} = \frac{4}{\pi}\frac{F}{DL}$$

Case where the Clearance cannot be Neglected

The force on an infinitesimal element of surface is:

$$dF(\theta) = P(\theta)dS = K\delta(\theta)dS = \tfrac{1}{2} \times K \times j \times \cos\theta / \cos\theta_0 - 1) \times dS$$

$$dF_y = \frac{Kj}{2}\left(\frac{\cos\theta}{\cos\theta_0} - 1\right)DL\cos\theta d\theta$$

$$F = \frac{KjDL}{2}\int_{-\theta_0}^{\theta_0}\left(\frac{\cos^2\theta}{\cos\theta_0} - \cos\theta\right)d\theta$$

$$= \frac{KjDL}{2}\left[\frac{\theta + \sin\theta\cos\theta}{2\cos\theta_0} - \sin\theta\right]_{-\theta_0}^{\theta_0}$$

thus,

$$F = \frac{KjDL}{2}\left(\frac{2\theta_0 + 2\sin\theta_0\cos\theta_0}{2\cos\theta_0} - 2\sin\theta_0\right) = \frac{KjDL}{2}\left(\frac{\theta_0 - \sin\theta_0\cos\theta_0}{\cos\theta_0}\right).$$

We recognise the trigonometric identity $\sin 2\theta = 2\sin\theta\cos\theta$:

$$F = \frac{KjDL}{4\cos\theta_0}(2\theta_0 - \sin 2\theta_0)$$

thus,

$$K = \frac{4F\cos\theta_0}{jDL(2\theta_0 - \sin 2\theta_0)}$$

and therefore:

$$P_{max} = Kj\frac{1-\cos\theta_0}{\cos\theta_0} = \frac{4F}{DL} \times \frac{1-\cos\theta_0}{2\theta_0 - \sin 2\theta_0}$$

Case of a Sphere-sphere Contact

A sphere-sphere contact corresponds to a spherical joint (socket/ball), such as a ball jointed cylinder saddle. It can also describe the situation of bearing balls.

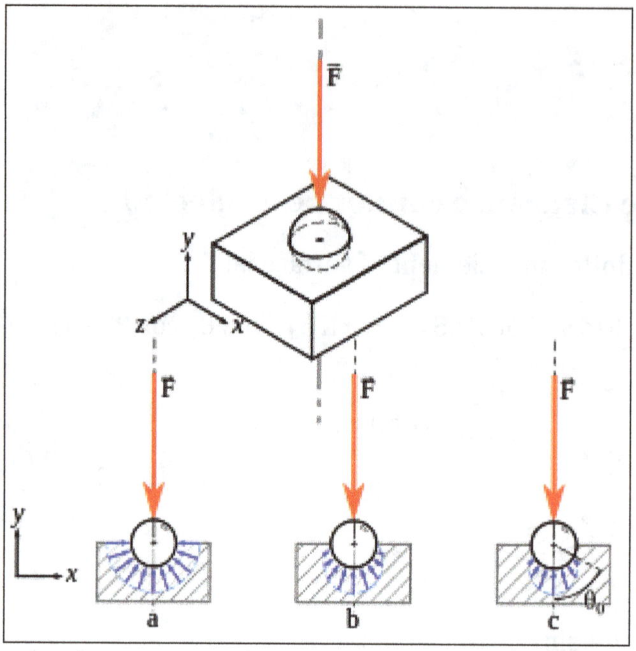

Bearing pressure in the case of a sphere-sphere contact.

Case of Uniform Pressure

The case is similar as above: when the parts are considered as rigid bodies and the clearance can be neglected, then the pressure is supposed to be uniform. It can also be calculated considering the projected area:

$$P = \frac{F}{\pi R^2} = \frac{\text{radial load}}{\text{projected area}}.$$

Case of a Sinusoidal Repartition of Pressure

As in the case of cylinder-cylinder contact, when the parts are modeled as elastic bodies with a negligible clearance, then the pressure can be modeled with a snusoidal repartition:

$$P(\theta, \varphi) = P_{max} \cdot \cos\theta$$

with

$$P = \frac{3}{2} \cdot \frac{F}{\pi R^2}$$

Hertz Contact Stress

When the clearance cannot be neglected, it is then necessary to know the value of the half contact angle θ_0, which cannot be determined in a simple way and must be measured. When this value is not available, the Hertz contact theory can be used.

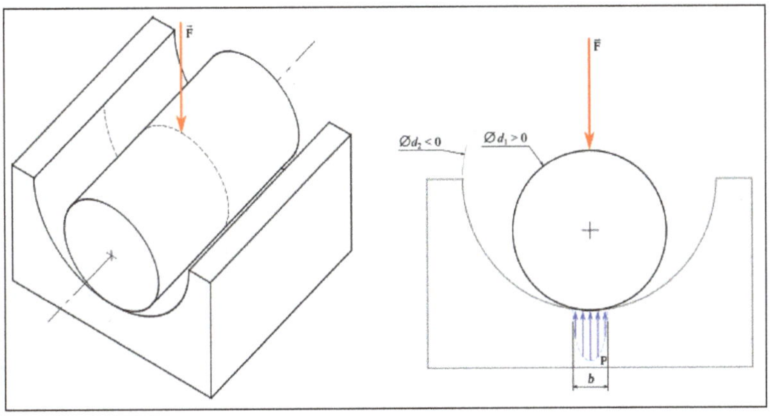

Hertz contact stress in the case of a male cylinder-female cylinder contact.

The Hertz theory is normally only valid when the surfaces cannot conform, or in other terms, cannot fit each other by elastic deformation; one surface must be convex, the other one must be also convex plane. This is not the case here, as the outer cylinder is concave, so the results must be considered with great care. The approximation is only valid when the inner radius of the container R_1 is far greater than the outer radius of the content R_2, in which case the surface container is then seen as flat by the content. However, in all cases, the pressure that is calculated with the Hertz theory is greater than the actual pressure (because the contact surface of the model is smaller than the real contact surface), which affords designers with a safety margin for their design. In this theory, the radius of the female part (concave) is negative.

A relative diameter of curvature is defined:

$$\frac{1}{d^*} = \frac{1}{d_1} + \frac{1}{d_2}$$

where d_1 is the diameter of the female part (negative) and d_2 is the diameter of the male part (positive). An equivalent module of elasticity is also defined:

$$\frac{1}{E^*} = \frac{1-\nu_1^2}{E_1} + \frac{1-\nu_2^2}{E_2}$$

where ν_i is the Poisson's ratio of the material of the part i and E_i its Young's modulus.

For a cylinder-cylinder contact, the width of the contact surface is:

$$b = \left(\frac{2F}{\pi L} \cdot \frac{d^*}{E^*} \right)^{1/2}$$

and the maximal pressure is in the middle:

$$P_{max} = \frac{2F}{\pi bL} = \sqrt{\frac{2FE^*}{\pi Ld^*}}.$$

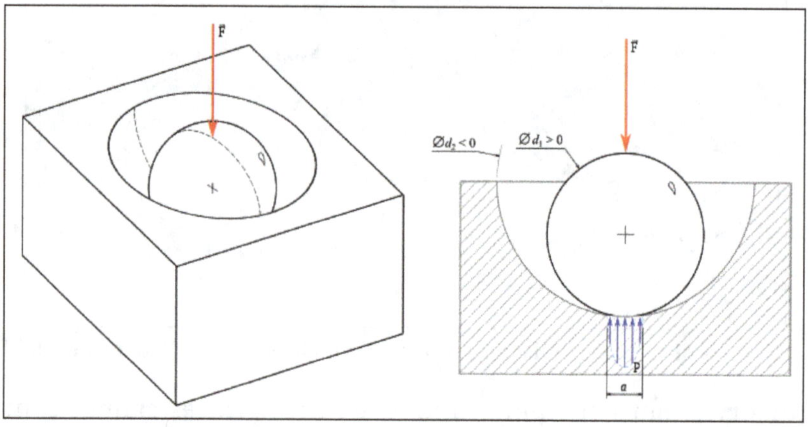

Hertz contact stress in the case of a male sphere-female sphere contact.

In case of a sphere-sphere contact, the contact surface is a disk whose radius is:

$$a = \left(\frac{3F}{8} \cdot \frac{d^*}{E^*}\right)^{1/3}$$

and the maximal pressure is in the middle:

$$P_{max} = \frac{3F}{\pi a^2} = \frac{4}{\pi}\sqrt[3]{3F\left(\frac{E^*}{d^*}\right)^2}$$

Applications

Bolt used as a Stop

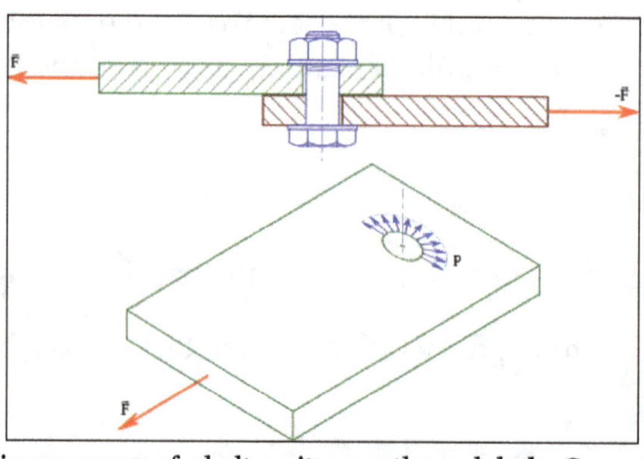

Bearing pressure of a bolt on its pass through hole. Case of two plates with a single overlap and one row of bolts.

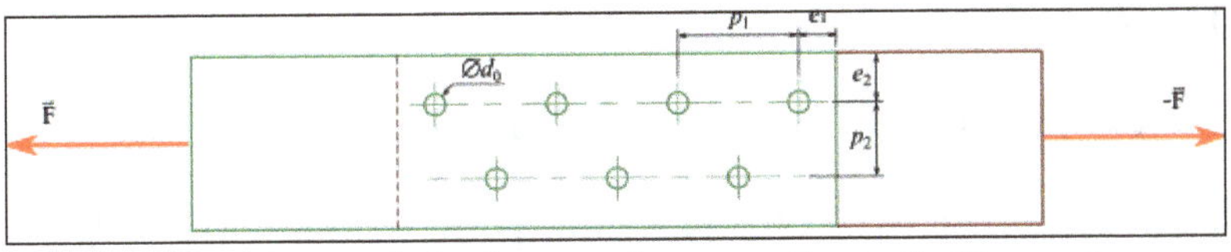

Dimensions used to design a bolted connection according to the Eurocode 3 standard.

In a bolted connection, the role of the bolts is normally to press one parts on the other; the adherence (friction) is opposed to the tangent forces and prevents the parts from sliding apart. In some cases however, the adherence is not sufficient. The bolts then play the role of stops: The screws endure shear stress whereas the hole endure bearing pressure.

In good design practice, the threaded part of the screw should be small and only the smooth part should be in contact with the plates; in the case of a shoulder screw, the clearance between the screw and the hole is very small (a case of rigid bodies with negligible clearance). If the acceptable pressure limit P_{lim} of the material is known, the thickness t of the part and the diameter d of the screw, then the maximum acceptable tangent force for one bolt $F_{b,Rd}$ (design bearing resistance per bolt) is:

$$F_{b,Rd} = P_{lim} \times d \times t.$$

In this case, the acceptable pressure limit is calculated from the ultimate tensile stress f_u and factors of safety, according to the Eurocode 3 standard. In the case of two plates with a single overlap and one row of bolts, the formula is:

$$P_{lim} = 1.5 \times f_u / \gamma_{M2}$$

where,

$\gamma_{M2} = 1.25$: partial safety factor.

In more complex situations, the formula is:

$$P_{lim} = k_1 \times \alpha \times f_u / \gamma_{M2}$$

where

- k1 and α are factors that take into account other failure modes than the bearing pressure overload; k1 take into account the effects that are perpendicular to the tangent force, and α the effects along the force.
- k1 = min{2.8e2/do ; 2.5} for end bolts, k1 = min{1.4p2/do ; 2.5} for inner bolts.
  - e2: Edge distance from the centre of a fastener hole to the adjacent edge of the part, measured at right angles to the direction of load transfer.
  - p2: Spacing measured perpendicular to the load transfer direction between adjacent lines of fasteners.

- do: Diameter of the passthrough hole.
- α = min{e1/3do ; p1/3do - 1/4 ; fub/fu ; 1}, with.
    - e1: End distance from the center of a fastener hole to the adjacent end of the part, measured in the direction of load transfer.
    - p1: Spacing between centers of fasteners in the direction of load transfer.
    - fub: Specified ultimate tensile strength of the bolt.

Table: Ultimate tensile stress for usual structural steels.

Steel grades (EN standard)	S235	S275	S355
Ultimate tensile stress f_u (MPa)	360	430	510

When the parts are in wood, the acceptable limit pressure is about 4 to 8.5 MP.

Plain Bearing

In plain bearings, the shaft is usually in contact with a bushing (sleeve or flanged) to reduce friction. When the rotation is slow and the load is radial, the model of uniform pressure can be used (small deformations and clearance).

The product of the bearing pressure times the circumferential sliding speed, called load factor PV, is an estimation of the resistance capacity of the material against the frictional heating.

Table: Acceptable bearing pressure.

Type of bushing Maximal circumferential sliding speed	Acceptable bearing pressure (MPa)
Self-lubricating bushels 7 to 8 m/s 13 m/s for graphite	graphite: 5 lead bronze: 20 to 30 tin bronze: 7 to 35
Composite bushing, Glacier 2 to 3 m/s	acetal: 70 PTFE: 50
Polymer bushing 2 to 3 m/s	7 to 10

Babbitt

Babbitt metal or bearing metal, is any of several alloys used for the bearing surface in a plain bearing.

The original Babbitt alloy was invented in 1839 by Isaac Babbitt in Taunton, Massachusetts, United States. He disclosed one of his alloy recipes but kept others as trade secrets. Other formulations were developed later. Like other terms whose eponymous origin is long since deemphasized (such as diesel engine or eustachian tube), the term babbitt metal is frequently styled in lowercase. It is preferred over the term "white metal", because the latter term may refer to various bearing alloys, lead- or tin-based alloys, or zinc die-casting metal.

Microstructure of Babbitt.

Babbitt metal is most commonly used as a thin surface layer in a complex, multi-metal structure, but its original use was as a cast-in-place bulk bearing material. Babbitt metal is characterized by its resistance to galling. Babbitt metal is soft and easily damaged, which suggests that it might be unsuitable for a bearing surface. However, its structure is made up of small hard crystals dispersed in a softer metal, which makes it a metal matrix composite. As the bearing wears, the softer metal erodes somewhat, which creates paths for lubricant between the hard high spots that provide the actual bearing surface. When tin is used as the softer metal, friction causes the tin to melt and function as a lubricant, which protects the bearing from wear when other lubricants are absent.

Internal combustion engines use Babbitt metal which is primarily tin-based because it can withstand cyclic loading. Lead-based Babbitt tends to work-harden and develop cracks but it is suitable for constant-turning tools such as sawblades.

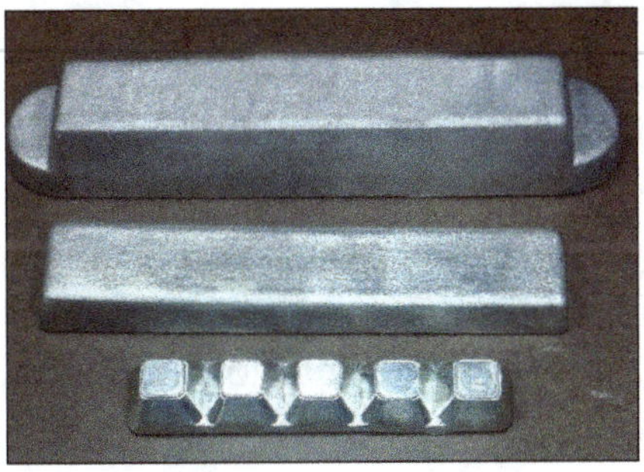

Pigs and bars of Grade 2 Babbitt.

Traditional Babbitt Bearings

In the traditional style of a babbitt metal bearing, a cast iron pillow block is assembled as a loose fit around the shaft, with the shaft in approximately its final position. The inner face of the cast iron pillow block is often drilled to form a key to locate the bearing metal as it is cast into place. The shaft is coated with soot as a release agent, the ends of the bearing are packed with clay, and molten metal poured into the cavity around the shaft, initially half filling the pillow block. The bearing is stripped, and the metal trimmed back to the top surface of the pillow block. Hardened babbitt metal is soft enough to be cut with a knife or sharp chisel.

A steel shim is inserted to protect the face of the lower bearing and to space the cap of the pillow block away from the shaft. After resealing the ends with clay, more metal is then poured to fill the cap of the pillow block through the hole in the top of the pillow block cap which will eventually be a lubrication hole.

The two halves of the bearing are then split at the shim, the oil holes cleared of metal and oil ways are cut into the surface of the new bearing. The shaft is smeared with engineer's blue and rotated in the bearing. When the bearing is disassembled the blue fills the hollows and is rubbed off the high spots. The high spots are scraped down, and the process repeated, until a uniform and evenly distributed pattern of blue shows when the shaft is removed. The bearing is then cleaned and lubricated, and shimmed up such that the shaft is held firmly but not binding in the bearing. The bearing is then "run in" by being run heavily lubricated at low load and revolution, completing the process of exposing the hard bearing surface. After final adjustment of the shimming, a very reliable and high load capability bearing results.

Before the advent of low cost electric motors, power was distributed through factories from a central engine via overhead shafts running in hundreds of Babbitt bearings.

The expression a "run bearing" also derives from this style of bearing, since failure of lubrication will lead to heat buildup due to friction in the bearing, eventually leading to the bearing metal liquefying and running out of the pillow block.

Modern Babbitt Bearings

Until the mid-1950s, poured Babbitt bearings were common in automotive applications. The Babbitt was poured into the block or caps using a form. Tin-based Babbitts were used as they could stand up to the impact loads found on the connecting rods and crankshaft. The poured Babbitt bearings were kept thin. The rods and caps would have shims that could be peeled off as the Babbitt wore. Ford was known to use two .002" on each cap and Babbitt that was 86% tin, 7% copper, 7% antimony. Steel shims were used as the brass shims used today tend to compress over time contributing to shorter bearing life. The poured Babbitt bearings commonly get over 50,000 miles of use before needing replacement. Poured Babbitt bearings are also known to fail gracefully allowing the car to be driven for extended periods of time. The failed bearing is not likely to damage the crankshaft.

In more modern practice, the crankshaft and connecting rod big end bearings in a modern automobile engine are made of a replaceable steel shell, keyed to the bearing caps. The inner surface of the steel shell is plated with a coating of bronze which is in turn coated with a thin layer of Babbitt metal as the bearing surface.

The process of laying down this layer of white metal is known as Babbitting.

Alternative Bearings

In many applications, rolling-element bearings, such as ball or roller bearings, have replaced Babbitt bearings. Though such bearings can offer a lower coefficient of friction than plain bearings, their key advantage is that they can operate reliably without a continuous pressurized supply of lubricant. Ball and roller bearings can also be used in configurations that are required to carry both radial and axial thrusts. However, rolling-element bearings lack the beneficial damping and shock-load capability provided by fluid-film bearings, such as the Babbitt.

Science of Bearing Babbitt Selection

The engineering of a bearing's Babbitt lining is usually completed during the design of the machine. In selecting the proper type of Babbitt for a particular job there are a number of factors to take into consideration, the most important of which are as follows:

- Surface speed of the shaft.
- Load that the bearing is required to carry.

There is no doubt that if a bearing is to be highly loaded in relation to its size, a high tin alloy is desirable; whereas for much slower speed work and less heavily loaded bearings, a lead-Based Babbitt may be employed, and is far more economical.

- Surface speed of the shaft: (The number of feet travelled per minute by the shaft circumferentially).

$$(\text{Pi} \times \text{D} \times \text{RPM}) / 12 = \text{S}$$

Example: Determine the surface of a 2 inch diameter shaft going 1,400 revolutions per minute (RPM).

$$(\text{Pi} \times \text{D} \times \text{RPM}) / 12 = (3.1416 \times 2 \times 1{,}400) / 12 = 733.04 \text{ Ft/min.}$$

Where: Pi = 3.1416, D = Diameter of shaft, S = Surface speed of the shaft.

- Load bearing is required to carry: (the weight which is being exerted through the combined weights of the shaft and any other direct weights on the shaft and measured in pounds per square inch).

$$\text{W} / (\text{I.D} \times \text{L.O.B.}) = \text{L.}$$

Example: Determine the load on a bearing of a 2-inch I.D bearing, 5 inches long and carrying a weight of 3,100 lbs.

W / (I.D x L.O.B.) = 3,100 / (2 x 5) = 310 Lbs/sq.in

Where: W = Total weight carried by bearing, I.D = Inside diameter of bearing, L.O.B = Length of Bearing, L = Load bearing required to carry.

Babbitt Bearing Selection

While not subject to precise calculations, the following considerations must also be taken into account:

- Continuity of service.
- Bonding characteristics.
- Cooling facilities.
- Lubrication.
- Cleanliness.
- Maintenance schedule for the bearing in use.
- For example, a bearing in continuous use in a harsh environment without regular maintenance will require different Babbitt and lubrication than a bearing in intermittent use in a clean, light duty environment. This so-called art is really the condensation of the experience of the technician and the experience of the bearing being rebuilt.

If the bearing has performed well in use over many years, the bearing needs simply to be rebuilt to its original specification and formulation. In this case the technician's greatest concerns are:

- Bearing shell surface preparation.
- Bonding characteristics of the tinning compound and the Babbitt layer.
- Load bearing surface preparation and finish.

Eco-babbitt

Eco-Babbitt is an alloy of 90% Sn, 7% Zn, 3% Cu that is not technically a Babbitt metal.

Bridge Bearing

A bridge bearing is a component of a bridge which typically provides a resting surface between bridge piers and the bridge deck. The purpose of a bearing is to allow controlled movement and thereby reduce the stresses involved. Possible causes of movement are thermal expansion and contraction, creep, shrinkage, or fatigue due to the properties of the material used for the bearing. External sources of movement include the settlement of the ground below, thermal expansion, and seismic activity. There are several different types of bridge bearings which are used depending on a number of different factors including the bridge span, loading conditions, and performance specifications. The oldest form of bridge bearing is simply two plates resting on top of each other. A common form of modern bridge bearing is the elastomeric bridge bearing. Another type of bridge bearing is the mechanical bridge bearing. There are several types of mechanical bridge bearing, such as the pinned bearing, which in turn includes specific types such as the rocker bearing, and the roller bearing. Another type of mechanical bearing is the fixed bearing, which allows rotation, but not other forms of movement.

An expansion bearing on the Queen Elizabeth II Metro Bridge.

The first bridge bearings to be used were plane bearings in the early 1800s, which included sliding bearings or roller bearings. Plane bearings allowed horizontal movement in one direction, and could therefore transfer horizontal load. Rotating bearings were used in the late 1800s and early 1900s and included rocker bearings, knuckle bearings, and ball bearings. Rotating bearings allowed movement in both the horizontal and vertical directions. Both plane bearings and roller bearings were made of metal. In the mid-1900s, deformation bearings began to be used, which were made of rubber. Deformation bearings primarily include elastomeric bearings, the most common type of bridge bearing used today.

Types of Bridge Bearings

- Rocker Bearings: Rocker bearings have curved surfaces that allow rocking. As the bridge expands, the bearing rocks to allow movement in the horizontal direction. Rocker bearings are primarily made of steel. Rocker bearings tend to be used for highway bridges.

- Elastomeric Bearings: Elastomeric bridge bearings are the most popular type of bridge bearing used today. They are made of rubber and do not have any moving parts, because the rubber itself allows movement in the bridge. Elastomeric bearings can be made at a low cost, and do not need to be maintained, like other forms of bearings that have moving parts and are made of metal. Elastomeric bearings can be reinforced with steel to make them stronger if needed.
- Sliding Bearings: Sliding bearings have both a flat sliding surface to allow horizontal movement and a spherical surface to allow rotation. Although they used to be made of metal, sliding bearings now tend to be made of Teflon.
- Spherical Bearings: As the name suggests, spherical bearings are in the shape of a sphere. These bearings only allow rotation, and prevent movement in the horizontal and vertical directions.

Bearing Materials

Bearings materials are classified as through-hardened materials (used largely for ball bearings) and case-hardened materials (used largely for roller bearings). Critical applications require vacuum-processed steels. Regardless of material type, a commonly accepted minimum hardness for bearing components is 58 R_c. Less hardness allows the races to Brinell. Since hardness decreases with increasing temperature, conventional bearing steels such as 440C and SAE 52100 cannot be used much above 350 °F.

Through-hardened materials have received much attention in an effort to develop steels suitable for high-temperature bearings. Alloys with various proportions of molybdenum, tungsten, chromium, vanadium, aluminum, and silicon are used to produce hardness at high temperatures.

The ball-bearing industry has used SAE 52100 steel as a standard material since 1920. This is a high-carbon chromium steel which also contains small amounts of manganese and silicon. This air-melted alloy is clean, hard, and wear resistant.

For corrosive environments, 440C stainless steel should be considered. However, bearings made of 440C do not have as high a dynamic capacity as those made of SAE 52100. Halmo and M-50 are usable up to about 600 °F. M-50 is the more widely used. T-1 and M-10 are good to 800 °F, M-1 and M-2 to 900 °F, and WB-49 to 1,000 °F. Oxidation resistances of M-1 and M-2 are marginal above 900 °F.

For M-50 and M-1 the dynamic capacity can be exceeded at 600 °F, but dynamic capacity of bearings made of WB-49 should be derated above that temperature. In addition to operating temperature, criteria for choosing a material should include fabrication costs. Generally, bearings made from M-series materials cost about 50% more than those made from SAE 52100, because of grinding difficulties. At temperatures below 350 °F, there appears to be no technical advantage of M-series materials over SAE 52100.

Carburized, or case-hardened materials, are characterized by a hardened surface (greater than 0.015 in. thick) and a soft core. At room temperatures, the surface hardness of these materials run around R_c 58 to 63, and their core hardnesses are R_c 25 to 48. These materials are generally limited to temperatures less than 350 °F. Two exceptions to the temperature rule are Timken CBS600 and CBS1000. If hot hardnesses of 58 R_c are required, CBS600 appears to be limited to 450 °F and CBS1000 to 600 °F.

No data are available to compare the dynamic capacities of carburized and through-hardened materials. Where shock and high vibrational loads are present, carburized materials may have an advantage because of their soft, ductile inner core.

Vacuum-melt steels provide increased reliability and basic dynamic load rating because they have a lower inclusion content than air-melted steels. However, the improvement in bearing life is not always commensurate with the improvement in cleanliness. Nonmetallic inclusions are present to some degree even in exceptionally clean steels and can be the nucleus of fatigue cracks.

Induction vacuum melting is a process in which a cold charge is melted in an induction furnace and then poured into ingots; the whole operation is performed in vacuum. Drawbacks of this process are high cost and variations in quality because of refractory problems. A better way to make bearing steels is consumable-electrode vacuum melting. Electrodes made from air-melted heat are remelted under vacuum by electric arcs. The remelted product solidifies in a water-cooled copper mold under vacuum. This gives different solidification conditions than the induction-melting technique and produces a more consistent high-quality steel.

Journal Bearings

Journal or plain bearings consist of a shaft or journal which rotates freely in a supporting metal sleeve or shell. There are no rolling elements in these bearings. Their design and construction may be relatively simple, but the theory and operation of these bearings can be complex.

Low-speed pins and bushings are a form of journal bearing in which the shaft or shell generally does not make a full rotation. The partial rotation at low speed, before typically reversing direction, does not allow for the formation of a full fluid film and thus metal-to-metal contact does occur within the bearing. Pins and bushings continually operate in the boundary lubrication regime.

These types of bearings are typically lubricated with an extreme pressure (EP) grease to aid in supporting the load. Solid molybdenum disulfide (moly) is included in the grease to enhance the load-carrying capability of the lubricant.

Many outdoor construction and mining equipment applications incorporate pins and bushings. Consequently, shock loading and water and dirt contamination are often major factors in their lubrication.

Kingsbury Radial and Thrust Pad Bearing.

Dry journal bearings consist of a shaft rotating in a dry sleeve, usually a polymer, which may be blended with solids such as molybdenum, graphite, PTFE or nylon. These bearings are limited to low-load and low-surface speed applications. Semilubricated journal bearings consist of a shaft rotating in a porous metal sleeve of sintered bronze or aluminum in which lubricating oil is contained within the pores of the porous metal. These bearings are restricted to low loads, low-to-medium velocity and temperatures up to 100 °C (210 °F).

Tilting-pad or pivoting-shoe bearings consist of a shaft rotating within a shell made up of curved pads. Each pad is able to pivot independently and align with the curvature of the shaft. A diagram of a tilt-pad bearing is presented in figure.

The advantage of this design is the more accurate alignment of the supporting shell to the rotating shaft and the increase in shaft stability which is obtained.

Journal bearings are meant to include sleeve, plain, shell and babbitt bearings. The term babbitt actually refers to the layers of softer metals (lead, tin and copper) which form the metal contact surface of the bearing shell. These softer metals overlay a stronger steel support shell and are needed to cushion the shell from the harder rotating shaft.

Simple shell-type journal bearings accept only radial loading, perpendicular to the shaft, generally due to the downward weight or load of the shaft. Thrust or axial loads, along the axis of the shaft, can also be accommodated by journal bearings designed for this purpose. Figure shows a tilt-pad bearing capable of accepting both radial and thrust loads.

Journal bearings operate in the boundary regime (metal-to-metal contact) only during the startup and shutdown of the equipment when the rotational speed of the shaft (journal) is insufficient to create an oil film. It is during startup and shutdown when almost all of the damage to the bearing occurs.

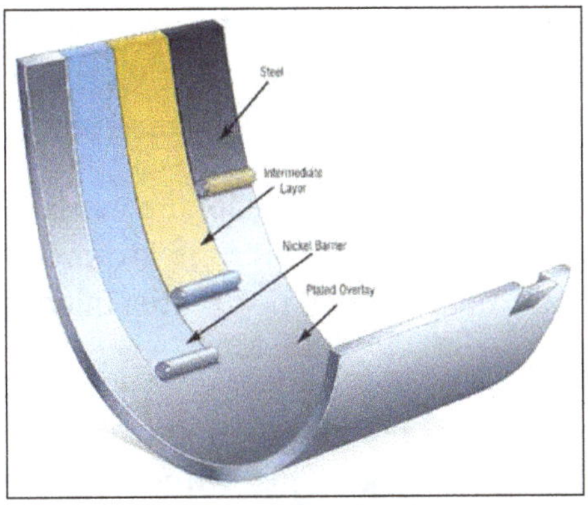

Layers of Journal Bearing Structure.

Hydrostatic lift, created by an external pressurized oil feed, may be employed to float large, heavy journals prior to startup (shaft rotation) to prevent this type of damage. During normal operation, the shaft rotates at sufficient speed to force oil between the conforming curved surfaces of the shaft and shell, thus creating an oil wedge and a hydrodynamic oil film.

This full hydrodynamic fluid film allows these bearings to support extremely heavy loads and operate at high rotational speeds. Surface speeds of 175 to 250 meters/second (30,000 to 50,000 feet/minute) are common. Temperatures are often limited by the lubricant used, as the lead and tin babbitt is capable of temperatures reaching 150 °C (300 °F).

It is important to understand that the rotating shaft is not centered in the bearing shell during normal operation. This offset distance is referred to as the eccentricity of the bearing and creates a unique location for the minimum oil film thickness.

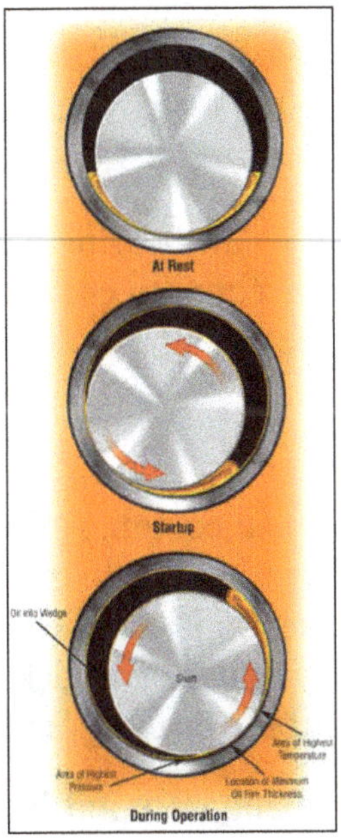

Shaft Motion during Startup.

Normally, the minimum oil film thickness is also the dynamic operating clearance of the bearing. Knowledge of the oil film thickness or dynamic clearances is also useful in determining filtration and metal surface finish requirements.

Typically, minimum oil film thicknesses in the load zone during operation ranges from 1.0 to 300 microns, but values of 5 to 75 microns are more common in midsized industrial equipment. The film thickness will be greater in equipment which has a larger diameter shaft.

The pressures encountered in the contact area of journal bearings are significantly less than those generated in rolling bearings. This is due to the larger contact area created by the conforming (similar curvature) surfaces of the journal and the shell.

The mean pressure in the load zone of a journal bearing is determined by the force per unit area or in this case, the weight or load supported by the bearing divided by the approximate load area of the bearing (the bearing diameter times the length of the bearing). In most industrial applications, these values range from 690 to 2,070 kPa (100 to 300 psi).

At these low pressures, there is virtually no increase in the oil viscosity in the bearing contact area due to pressure. Automotive reciprocating engine bearings and some severely loaded industrial applications may have mean pressures of 20.7 to 35 MPa (3,000 to 5,000 psi). At these pressure levels, the viscosity may slightly increase. The maximum pressure encountered by the bearing is typically about twice the mean value, to a maximum of about 70 MPa (10,000 psi).

Oil whirl is a phenomenon that can occur in high-speed journal bearings when the shaft position within the shell becomes unstable and the shaft continues to change its position during normal operation, due to the fluid forces created within the bearing. Oil whirl may be reduced by increasing the load or changing the viscosity, temperature or oil pressure in the bearing.

A permanent solution may involve a new bearing with different clearances or design. Oil whip occurs when the oil whirl frequency coincides with the system's natural frequency. The result can be a catastrophic failure.

Oil Lubrication

Oils are used in journal bearings when cooling is required or contaminants or debris need to be flushed away from the bearing. High-speed journal bearings are always lubricated with oil rather than a grease. Oil is supplied to the bearing by either a pressurized oil pump system, an oil ring or collar or a wick. Grooves in the bearing shell are used to distribute the oil throughout the bearings' surfaces.

The viscosity grade required is dependent upon bearing RPM, oil temperature and load. The bearing speed is often measured strictly by the revolutions per minute of the shaft, with no consideration of the surface speed of the shaft, as per the "nd_m" values calculated for rolling bearings. Table provides a general guideline to selecting the correct ISO viscosity grade.

The ISO grade number indicated is the preferred grade for speed and temperature range. ISO 68- and 100-grade oils are commonly used in indoor, heated applications, with 32-grade oils being used for high-speed (10,000 RPM) units and some outdoor low-temperature applications.

The table that the higher the bearing speed, the lower the oil viscosity required; and that the higher the operating temperature of the unit, the higher the oil viscosity that is required. If vibration or minor shock loading is possible, a higher grade of oil than the one indicated in table should be considered.

Table: Journal Bearing ISO Viscosity Grade Selection.

Bearing Speed	Bearing / Oil Temperature (°C)			
(rpm)	0 to 50	60	75	90
300 to 1,500	-	68	100 to 150	-
~1,800	32	32 to 46	68 to 100	100
~3,600	32	32	46 to 68	68 to 100
~10,000	32	32	32	32 to 46

Another method of determining the proper viscosity grade is by applying minimum and optimum viscosity criteria to a viscosity-temperature plot. A generally accepted minimum viscosity of the oil at the operating temperature for journal bearings is 13 cSt, although some designs allow for an oil as thin as 7 or 8 cSt at the operating temperature.

The optimum viscosity at operating temperature is 22 to 35 cSt, for moderate-speed bearings if no shock loading occurs. The optimum viscosity may be as high as 95 cSt for low-speed, heavily loaded or shock-loaded journal bearings.

Using this method requires some knowledge of the oil temperature within the bearing under operating conditions, which can be difficult to determine. Fortunately, an accurate oil temperature is not needed for most viscosity determinations. It is common to determine the temperature of the outer surface of the pipes carrying oil to and away from the bearing.

The temperature of the oil inside of the pipes will generally be higher (5 to 10 °C, 10 to 18 °F) than the outer metal surface of the pipe. The oil temperature within the bearing can be taken as the average of the oil entering versus the temperature exiting the bearing.

A third and more complex method is to calculate the oil viscosity needed to obtain a satisfactory oil film thickness. Persons wishing to learn more about this method should seek information regarding the Sommerfeld equation and either eccentricity ratios or Reynolds Numbers. If the oil selected is too low in viscosity, heat will generate due to an insufficient film thickness and some metal-to-metal contact will occur. If the oil is too high in viscosity, heat will again be generated, but due to the internal fluid friction created within the oil. Selecting an oil which is too high in viscosity can also increase the likelihood of cavitation.

The high- and low-pressure zones, which are created within the oil on each side of the area of minimum film thickness, can cause oil cavitation in these bearings. Cavitation is a result of expansion of dissolved air or a vapor (water or fuel) in the low-pressure zone of the bearing.

The resulting bubble implodes, causing damage, as it passes through the high-pressure portion of the bearing. If the implosion or collapse of the vapor bubble occurs next to the metal surface, this can cause cavitation pitting damage to the metal. If the implosion of the bubble occurs within the oil, a micro hot spot or micro-dieseling can occur, which may lead to varnishing within the system.

Typically, a rust and oxidation (R&O) inhibited additive system is used in the oils employed in these applications. Antifoam and pour point depressant additives may also be present. Antiwear (AW) hydraulic oils may also be used as long as the high-temperature limit of the zinc AW component is not exceeded and excessive water is not present. R&O oils tend to have better water separation characteristics which is beneficial, and the AW properties of a hydraulic oil would be beneficial only during startup and shutdown, assuming a properly operating bearing.

Grease Lubrication

Grease is used to lubricate journal bearings when cooling of the bearing is not a factor, typically if the bearing operates at relatively low speeds. Grease is also beneficial if shock loading occurs or if the bearing frequently starts and stops or reverses direction.

Grease is almost always used to lubricate pins and bushings because it provides a thicker lubricant than oil to support static loads and to protect against vibration and shock-loading that are common in many of these applications.

Lithium soap or lithium complex thickeners are the most common thickeners used in greases and are excellent for most journal bearing applications. The grade of grease used is typically an NLGI grade 2 with a base oil viscosity of approximately 150 to 220 cSt at 40 °C.

Greases for low-speed, high-load, high temperatures and for pins and bushings may use a higher viscosity base oil and be formulated with EP and solid additives. Greases for improved water resistance may be formulated with heavier base oils, different thickeners and special additive formulations. Greases for better low-temperature dispensing may incorporate a lower viscosity base oil manufactured to an NLGI 1 specification. Bearings lubricated by a centralized grease dispensing systems typically use a 1, 0 or 00 grade of grease.

The apparent viscosity of grease changes with shear (pressure, load and speed) that is, greases are non-Newtonian or thixotropic. Within a rotating journal bearing, as the bearing rotates faster (shear rate increases), the apparent viscosity of the grease decreases and approaches the viscosity of the base oil used in grease. At both ends of the bearing shell, the pressure is lower and therefore the apparent viscosity remains higher. The resulting thicker grease at the bearing ends acts as a built-in seal to reduce the ingression of contaminants.

Greasing Procedures

The greasing procedures for journal bearings and pins and bushings are not as well-defined or as critical as for rolling bearings because the grease is not subjected to the churning action created by the rolling elements.

The volume of grease to inject and the frequency of application are dictated more by trial and error. Generally, most journal bearings cannot be overgreased. Caution must be taken when pumping grease into a bearing that is fitted with seals, so they are not damaged or displaced by the force and volume of the incoming grease. The harshness of the environment, shock loading and especially the operating temperature will be major factors in determining the frequency of relubrication.

Journal bearings are generally a simpler design and not as difficult to lubricate as rolling element bearings. The proper viscosity matched to the operating conditions and a clean and dry lubricant will usually suffice to form a full fluid lubricating film and provide excellent bearing life.

Air Bearing

Air bearings (also known as aerostatic or aerodynamic bearings) are bearings that use a thin film of pressurized gas to provide a low friction load-bearing interface between surfaces. The two surfaces do not touch, thus avoiding the traditional bearing-related problems of friction, wear, particulates, and lubricant handling, and offer distinct advantages in precision positioning, such as lacking backlash and static friction, as well as in high-speed applications.

A differentiation is made between aerodynamic bearings, which establish the air cushion through the relative motion between static and moving parts, and aerostatic bearings, in which the pressure is being externally inserted.

Gas bearings are mainly used in precision machinery tools (measuring and processing machines) and high-speed machines (spindle, small-scale turbomachinery, precision gyroscopes).

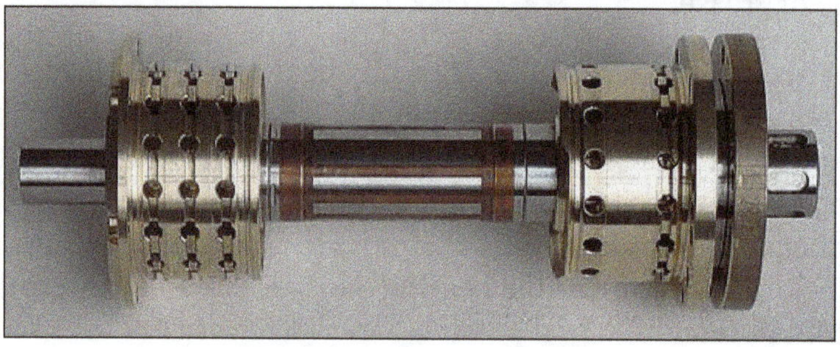

Air borne high frequency spindle.

Gas Bearing Types

Gas-lubricated bearings are classified in two groups, depending on the source of pressurization of the gas film providing the load-carrying capacity:

- Aerostatic bearings: The gas is externally-pressurized (using a compressor or a pressure tank) and injected in the clearance of the bearing. Consequently, aerostatics bearings can sustain a load even in absence of relative motion but require an external gas compression system, which induces costs in terms of complexity and energy.

- Aerodynamic bearings: The gas is pressurized by the relative velocity between the static and moving surfaces in the bearing. Such bearings are self-acting and do not require an external input of compressed gas. However, mechanical contact occurs at zero speed, requiring a particular tribological consideration to avoid premature wear.

Hybrid bearings combining the two families also exist. In such cases, a bearing is typically fed with externally-compressed gas at low speed and then relies partially or entirely on the self-pressurizing effect at higher speeds.

Among these two technological categories, gas bearings are classified depending on the kind of linkage they realize:

- Linear-motion bearings: Support a translation along 1 or 2 directions between two planes.

- Journal bearings: Support a rotation between two parts.

- Thrust bearings: Block the axial displacement of a rotating part, usually used in combination with journal bearings.

The main air bearing types fall under the following categories:

Gas Bearing Type	Technology	Description
Aerostatic	Porous Media	Gas flow is controlled through porous material.
	Micro-Nozzle	Gas flow is controlled through micro-sized holes.
	Orifice Type	Gas flow is controlled through holes and grooves.
	Air caster	Gas flow is controlled through an air bag.
Aerodynamic	Foil bearing	Bearing surface is flexible, allowing large displacement and providing a good stability.
	Spiral groove bearing	Gas film is pressurized by grooves machined on one of the surfaces, achieving high load capacity and stability. The usual groove patterns are herringbone-shaped, spiral or straight (step bearings).

Aerostatic Bearings

Pressurized gas acts as a lubricant in the gap between bearing moving parts. The gas cushion carries the load without any contact between the moving parts. Normally, the compressed gas is supplied by a compressor. A key goal of supplying the gas pressure in the gap is that the stiffness and damping of the gas cushion reaches the highest possible level. In addition, gas consumption and uniformity of gas supply into the gap are crucial for the behaviors of aerostatic bearings.

Delivery of Gas to the Gap

Supplying gas to the interface between moving elements of an aerostatic bearing can be achieved in a few different methods:

- Porous Surface,
- Partial porous surface,
- Discrete orifice feeding,
- Slot feeding,
- Groove feeding.

There is no single best approach to feeding the film. All methods have their advantages and disadvantages specific to each application.

Dead Volume

Dead volumes refer in particular to chambers and canals existing in conventional aerostatic bearings in order to distribute the gas and increase the compressed pressure within the gap. The cavity inside porous (sintered) gas bearings are also attributed to dead volume.

Conventional Aerostatic Bearings

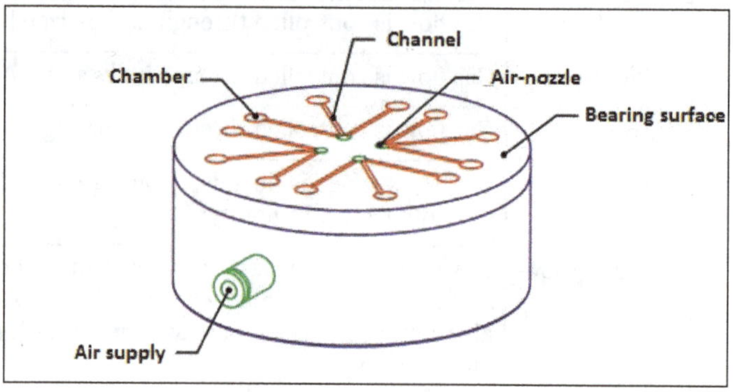

Nozzle-air bearing with chambers and channels.

With conventional single nozzle aerostatic bearings, the compressed air flows through a few relatively large nozzles (diameter 0.1 – 0.5 mm) into the bearing gap. The gas consumption thus allows only some flexibility such that the bearing's features (force, moments, bearing surface, bearing gap height, damping) can be adjusted only insufficiently. However, in order to allow a uniform gas pressure even with only some nozzles, aerostatic bearing manufacturers take constructive techniques. In doing so, these bearings cause dead volumes (non-compressible and thus weak air volume). In effect, this dead volume is very harmful for the gas bearing's dynamic and causes self-excited vibrations.

Single-nozzle Aerostatic Bearings

The pre-pressured chamber consists of a chamber around the centralized nozzle. Usually, this chamber's ratio is between 3% and 20% of the bearing's surface. Even with a chamber depth of 1/100 mm, the dead volume is very high. In the worst cases, these air bearings consist of a concave bearing surface instead of a chamber. Disadvantages of these air bearings include a very poor tilt stiffness.

Gas Bearings with Channels and Chambers

Typically, conventional aerostatic bearings are implemented with chambers and canals. This design assumes that with a limited amount of nozzles, the dead volume should decrease while distributing the gas within the gap uniformly. Most constructive ideas refer to special canal structures. Since the late 1980s, aerostatic bearings with micro canal structures without chambers are manufactured. However, this technique also has to manage problems with dead volume. With an increasing gap height, the micro canal's load and stiffness decreases. As in the case of high-speed linear drives or high-frequency spindles, this may cause serious disadvantages.

Laser Drilled Micro-nozzle Aerostatic Bearings

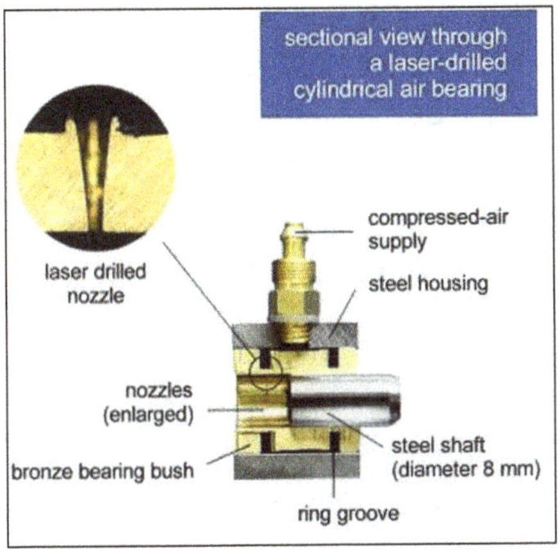

Cut through a cylindrical element.

Laserprocessing (cut through a bearing element).

Laser-drilled micro nozzle aerostatic bearings make use of computerized manufacturing and design techniques to optimize performance and efficiency. This technology allows manufacturers more flexibility in manufacturing. In turn this allows a larger design envelope in which to optimize their designs for a given application. In many cases engineers can create air bearings that approach the theoretical limit of performance. Rather than a few large nozzles, aerostatic bearings with lots of micro nozzles avoid dynamically disadvantageous dead volumes. Dead volumes refer to all cavities in which gas cannot be compressed during decrease of the gap. These appear as weak gas pressure stimulates vibration. Examples of the benefits are: linear drives with accelerations of more than 1,000 m/s^2 (100 g), or impact drives with even more than 100,000 m/s^2 (10,000 g) due to high damping in combination with dynamic stiffness; sub-nanometer movements due to lowest noise-induced errors; and seal-free transmission of gas or vacuum for rotary and linear drives via the gap due to guided air supply.

Micro-nozzle aerostatic bearings achieve an effective, nearly perfect pressure distribution within the gap with a large number of micro nozzles. Their typical diameter is between 0.02 mm and 0.06 mm. The narrowest cross-section of these nozzles lies exactly at the bearing's surface. Thereby the technology avoids a dead volume on the supporting air bearing's surface and within the area of the air supplying nozzles.

The micro nozzles are automatically drilled with a laser beam that provides top-quality and repeatability. The physical behaviors of the air bearings prove to have a low variation for large as well as for small production volumes. In contrast to conventional bearings, with this technique the air bearings require no manual or costly manufacturing.

The advantages of the micro-nozzle air bearing technology include:

- Efficient use of the air cushion (close to the physical limit) through a uniform pressure within the whole gap.

- Perfect combination of static and dynamic properties.
- Highest-possible flexibility of the air bearing properties with a particular gap height, it is possible to optimize the air bearing such that it has, for example, a maximum load, stiffness, tilt stiffness, damping, or a minimum air consumption (respectively also in combination with others).
- Multi-approved highest precision of all air bearings, e.g. In the measurement technology due to slightest movements (<< 2 nanometer) through physical, lowest-possible self-excited vibrations.
- Considerably higher tilt stiffness than conventional air bearings such that the air within the gap flows through canals from the loaded to the unloaded areas away.
- Vibration-free within the entire operating range even with high air pressure supply (actually even much more than 10 bar are possible).
- Highest reliability due to the large number of nozzles: Clogging of nozzles by particles is out of question (no failure in operation) because their diameters are much higher than the gap height.
- Possibility to adjust bearing properties for deformation and tolerances of the bearing and opposite surface.
- Proven usability for many bearing materials and coatings.

Some of these advantages, such as the high flexibility, the excellent static and dynamic properties in combination, and a low noise excitation, prove to be unique among all other aerostatic bearings.

Various Designs

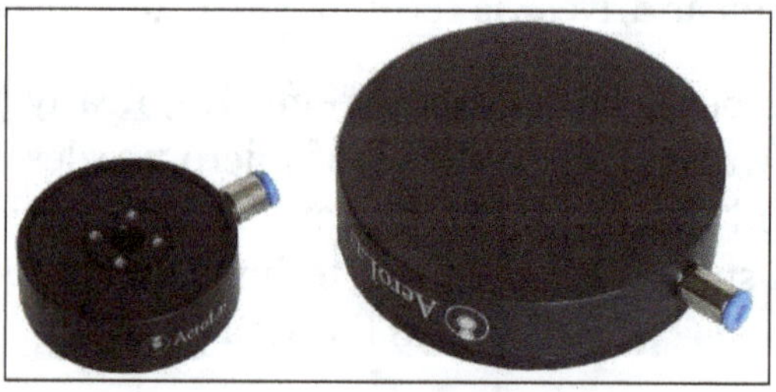

Standard bearing round.

Standard air bearings are offered with various mountings to link them in a system:

- Bearings for flexible connection with ball-pins. This design for standard air bearings is usually supplied on the market.
- Bearings with a high-stiff joint instead of a conventional ball-pin. Using this version the stiffness of the complete system is significantly higher.
- Bearings with integrated piston for preload of statically determined guidances.

- In addition, there are also rectangular bearings with a fixed mounting (joint-less) for guidances with highest stiffness for highest accuracy or highest dynamic.
- Furthermore, there are also air bearings with integrated vacuum or magnetic preloads, air bearings for high temperatures with more than 400 °C, as well as ones manufactured with alternative materials.

Advantages and Disadvantages of Gas-lubricated Bearings

Advantages

- Wearless operation, durability: Air bearings operate contact-free and so without abrasion. The only friction results from airflow between the bearing surfaces. Thus, the durability of air bearings is unlimited if they are designed and calculated correctly. Roller bearings and friction bearings have a high degree of friction when used at high speed or acceleration, causing a positive feedback loop where high abrasion decreases precision, which in turn causes greater wear, leading to their eventual failure.
- Guiding, repeatability, and position accuracy: In the chip production and when positioning at the back-end, repeatability accuracy of 1-2 μm must be reached with the wire bonder. At the die bonder, even 5 μm must be achieved. With such a precision, roller bearings reach their physical limit without a lower acceleration. At the front end (lithography), air bearings are already established.
- Cost advantage and repeatability: When applied in series, gas bearings can have a cost advantage over roller bearings: the production of a roller-guided high-frequency spindle is – according to a manufacturer – about 20% more expensive than air-guided spindles.
- Environmental purity: Because they do not require the use of oil for their lubrication and are frictionless, gas bearings are suited for applications requiring a low contamination of the working fluid. This is a critical aspect to the pharmaceutical industry, nuclear fuel processing, semi-conductor manufacturing and energy conversion cycles.

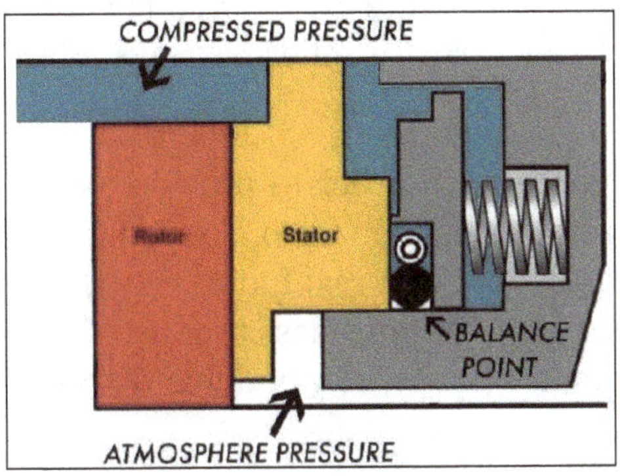

Gas-lubricated Bearings.

Disadvantages

- Self-excited vibration: In journal bearings, self-excited vibration can appear past a given speed, because of the cross-coupled stiffness and low damping of gas lubrication. This vibration can lead to an instability and threaten the gas bearing operation. Precise dynamic computations are required to ensure a safe operation within the desired speed range. This kind of instability is known as "half-speed whirl" and affects particularly aerodynamic bearings.

- Tight manufacturing tolerances: In order to carry sufficient load and avoid the instability mentioned above, tight tolerances are required in the clearance between bearing surfaces. Typical clearances ranging from 5 μm to 50 μm are required for both aerodynamic and aerostatic bearings. Consequently, air bearings are expensive to manufacture.

- Clean environment: Because of their small clearance, gas-lubricated bearings are sensitive to the presence of particulates and dust in the environment (in the case of aerodynamic bearings) and externally-pressurized gas (aerostatic bearings).

Theoretical Modeling

Gas-lubricated bearings are usually modeled using the Reynolds equation to describe the evolution of pressure in the thin film domain. Unlike liquid-lubricated bearings, the gas lubricant has to be considered as compressible, leading to a non-linear differential equation to be solved. Numerical methods such as Finite difference method or Finite element method are common for the discretization and the resolution of the equation, accounting for the boundary conditions associated to each bearing geometry (linear-motion, journal and thrust bearings). In most cases, the gas film can be considered as isothermal and respecting the ideal gas law, leading to a simplification of the Reynolds equation.

Automotive Technology

- Air-guided high-frequency knife drive: Even for movements which cause damage due to disruptive wear with roller bearings, lifetimes of the drive systems are unlimited.

- Air-guided turbo charger: In order to provide confidence and for the first investigations, an initial conversion from a conventional oil-guided turbo charger into air-guided was done. For a real future version, the use of results obtained from high-temperature solutions, mass products (proved production costs) and high-frequency spindles (know-how of dynamic background) will be very helpful.

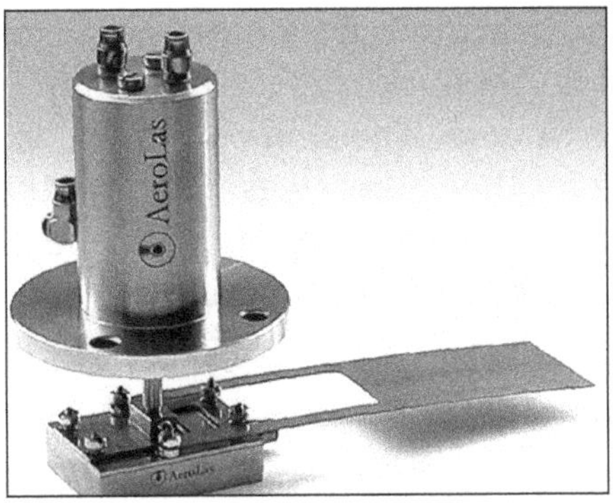

Bearing with piston actor.

Semiconductor Technology

- Air bearing for inspection device: In terms of the measurement of wafers and flat panels, it is very important to place the sensor chip precisely and without any contact along the surface. Therefore, the chip is integrated directly into the bearing's surface. The maximum distance tolerance to the surface which refers to the gap variation of the air bearing, is smaller than 0.5 µm. When placing the air bearing with the sensor chip, they must not touch the wafer surface being measured. As for the up-and-down movement, a pneumatic piston is used which is, for repeatability reasons, also air-guided. The preload of the air bearing and thus the gap height are also adjusted with this piston.

- Chuck with integrated lift drive: For the electrical testing of wafers the chuck can be lifted stick-slip-free up to 3 mm. The needed contact force for the probe is adjustable and independent from stroke. The lift drive is based on a voice coil motor; the guidance is air-guided. An air-guided pneumatic piston between the chuck and the drive limits the contact force.

Linear Drives

- Precision measurement stage: The filigree structure enables through light measurements for the 300 nm chip production with the utmost precision of less than 1 nm. In particular, the air bearings are designed for lowest air consumption with the highest stiffness.

- High-accelerated Doppler drive: The High-accelerated Doppler drive supports and guides a carbon fiber mirror (surface 500 mm x 250 mm) with an acceleration of up to 300 m/s² and a flexible movement profile with high precision. The solution consists of an air-guided drive: The beam (length 900 mm), which is fixed at the mirror, is manufactured of carbon fibre and carries the magnets of the linear motors. Cables/tubes (motor, air bearing, measurement system) do not move in order to avoid breakages due to high load cycles. The air-bearings are absolutely insensitive against geometric fluctuation as a result of a temperature change.

- Drive for production machine: Beside the performance, the reliability is extremely important for a production machine. The air-guided solution is designed to be statically determined. The iron-core linear motor and piston bearings achieve the preload for the air bearings. Thereby, the drive is easy to assemble and insensitive against geometric variations, for instance through temperature influences or the disposition of the machines.

Medical Technology

Fat- and oil-free drives for respirators, stick-slip-free movements of scanners or a high rotary speed of large rotors have all been achieved with air bearings.

- Air-guided computed tomography: High rotary speed (> 5.5 Hz / 330 rpm), low operation costs, no noise, large inner rotor diameter (> 1 m), small weight of rotor and frame, tilt possibility of the rotor as well as a high reliability. Besides a direct drive, a belt drive is also possible.

Production Technology

Primarily, stick-slip-free movements and/or smallest forces are required. The air bearing technology is predestinated for fat/oil-free high-dynamic movements with short strokes.

- Air bearing for the adjustment of components: With air-guided units, optical components can be arranged to have the same diameter on a rotary table. The air bearing with vacuum preload and a constant bearing gap height floats contact-less on top of the rotary table.

- Adjustment slider for optics production: The linear slider, which is air-guided and statically determined, guarantees a high-precision positioning of the optical component before grinding. The self-aligning process is done without friction or force. When clamped the component retains its position for further manufacturing in the sub-micrometer-range.

Space Technology

- Air-magnetic slip system: When transporting solar panels for satellites in a launching rocket, these must be folded. After reaching orbit, they unfold via a spring mechanism, weightlessly and without friction. This process requires prior testing on Earth due to reliability reasons. During the testing design, the solar panels are hung on magnetic preloaded air-bearings that compensate for gravity. In doing so, the unfolding movement process is carried out with a minimum friction impact which means that the solar panels are tested at close to reality. Moreover, the design offers absolutely maintenance-free handling with equal sequential movements.

- The air-bearing components (diameter 34 mm) with integrated magnets are so small such that they are able to glide contact-free along conventional rolled sheet plates smoothly and with a bearing gap height of about 25 μm. The holding force of an air bearing for one solar panel averages 600 N. This force is achieved by an equal distribution of the load on 16 single air bearing elements. The unfolding process of the solar panels has been developed for an area of 21 m x 2.5 m. The permanent magnetic preloaded air-bearing guidance system may be used for many types of hanging transportation movements as well as for many other applications, such as for instance for the stick-slip-free positioning of components during assembly.

Foil Bearing

A foil-air bearing for the core rotor shaft of an aircraft turbine engine.

A foil bearing, also known as a foil-air bearing, is a type of air bearing. A shaft is supported by a compliant, spring-loaded foil journal lining. Once the shaft is spinning fast enough, the working fluid (usually air) pushes the foil away from the shaft so that no contact occurs. The shaft and foil are separated by the air's high pressure, which is generated by the rotation that pulls gas into the bearing via viscosity effects. A high speed of the shaft with respect to the foil is required to initiate the air gap, and once this has been achieved, no wear occurs. Unlike aerostatic or hydrostatic bearings, foil bearings require no external pressurisation system for the working fluid, so the hydrodynamic bearing is self-starting.

Development

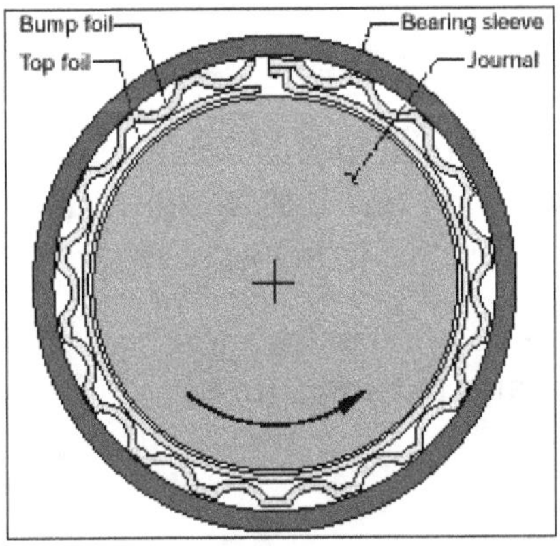

Foil Bearing.

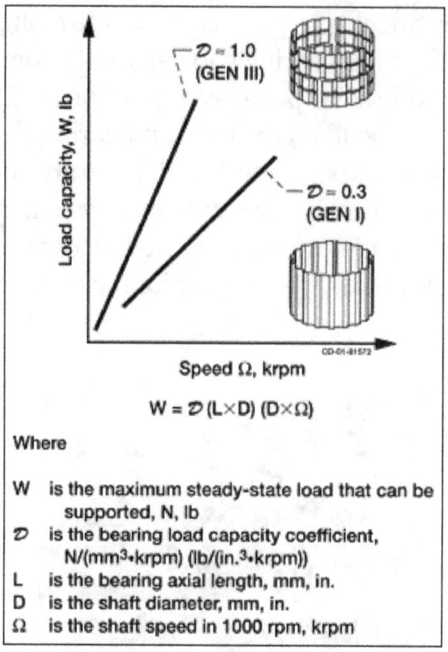

Load capacity against rotation speed, for Gen I and Gen III bearings.

Foil bearings were first developed in the late 1950s by AiResearch Mfg. Co. of the Garrett Corporation using independent R&D funds to serve military and space applications. They were first tested for commercial use in United Airlines Boeing 727 and Boeing 737 cooling turbines in the early and mid-1960s. Garrett AiResearch air cycle machine foil bearings were first installed as original equipment in 1969 in the DC-10's environmental control systems. Garrett AiResearch foil bearings were installed on all US military aircraft to replace existing oil-lubricated rolling-contact bearings. The ability to operate at cryogenic gas temperatures and at very high temperatures gave foil bearings many other potential applications.

Current-generation foil bearings with advanced coatings have greatly exceeded the limitations of earlier designs. Antiwear coatings exist that allow over 100,000 start/stop cycles for typical applications.

Applications

Turbomachinery is the most common application because foil bearings operate at high speed. The main advantage of foil bearings is the elimination of the oil systems required by traditional bearing designs. Other advantages are:

- Higher efficiency, due to a lower heat loss to friction; instead of fluid friction, the main source of heat is parasitic drag.
- Increased reliability.
- Higher speed capability.

- Quieter operation.
- Wider operating temperature range (40–2,500 K).
- High vibration and shock load capacity.
- No scheduled maintenance.
- No external support system.
- Truly oil free where contamination is an issue.
- Capable of operating above critical speed.

Areas of current research are:

- Higher load capacity.
- Improved damping.
- Improved coatings.

The main disadvantages are:

- Lower capacity than roller or oil bearings.
- Wear during start-up and stopping.
- High speed required for operation.

Linear-motion Bearing

A linear-motion bearing or linear slide is a bearing designed to provide free motion in one direction. There are many different types of linear motion bearings.

Motorized linear slides such as machine slides, X-Y tables, roller tables and some dovetail slides are bearings moved by drive mechanisms. Not all linear slides are motorized, and non-motorized dovetail slides, ball bearing slides and roller slides provide low-friction linear movement for equipment powered by inertia or by hand. All linear slides provide linear motion based on bearings, whether they are ball bearings, dovetail bearings, linear roller bearings, magnetic or fluid bearings. X-Y tables, linear stages, machine slides and other advanced slides use linear motion bearings to provide movement along both X and Y multiple axis.

Rolling-element Bearing

A rolling-element bearing is generally composed of a sleeve-like outer ring and several rows of balls retained by cages. The cages were originally machined from solid metal and were quickly replaced by stampings. It features smooth motion, low friction, high rigidity and long life. They are economical, and easy to maintain and replace.

- Rolling-element bearings can only run on hardened steel or stainless steel shafting (raceways).
- Rolling-element bearings are more rigid than plain bearings.
- Rolling-element bearings do not handle contamination well and require seals.
- Rolling-element bearings require lubrication.

Rolling-element bearings are manufactured in two forms: ball bearing slides and roller slides.

Ball Bearing Slides

Also called "ball slides," ball bearing slides are the most common type of linear slide. Ball bearing slides offer smooth precision motion along a single-axis linear design, aided by ball bearings housed in the linear base, with self-lubrication properties that increase reliability. Ball bearing slide applications include delicate instrumentation, robotic assembly, cabinetry, high-end appliances and clean room environments, which primarily serve the manufacturing industry but also the furniture, electronics and construction industries. For example, a widely used ball bearing slide in the furniture industry is a ball bearing drawer slide.

Commonly constructed from materials such as aluminum, hardened cold rolled steel and galvanized steel, ball bearing slides consist of two linear rows of ball bearings contained by four rods and located on differing sides of the base, which support the carriage for smooth linear movement along the ball bearings. This low-friction linear movement can be powered by either a drive mechanism, inertia or by hand. Ball bearing slides tend to have a lower load capacity for their size compared to other linear slides because the balls are less resistant to wear and abrasions. In addition, ball bearing slides are limited by the need to fit into housing or drive systems.

The travelling distance of linear recirculating ball bearings is only limited by the length of their rail, as the balls recirculate inside the bearing's housing. Linear non-recirculating ball bearings have balls installed on a bracket and only move in one axis without recirculation. Since the balls do not recirculate, this type of bearings can provide extremely smooth motion. However, the travelling distance of linear non-recirculating ball bearings is limited by the length of the bracket.

Roller Slides

Also known as crossed roller slides, roller slides are non-motorized linear slides that provide low-friction linear movement for equipment powered by inertia or by hand. Roller slides are based on linear roller bearings, which are frequently criss-crossed to provide heavier load capabilities and better movement control. Serving industries such as manufacturing, photonics, medical and telecommunications, roller slides are versatile and can be adjusted to meet numerous applications which typically include clean rooms, vacuum environments, material handling and automation machinery.

Roller slides work similarly to ball bearing slides, except that the bearings housed within the carriage are cylinder-shaped instead of ball shaped. The rollers crisscross each other at a 90° angle and move between the four semi-flat and parallel rods that surround the rollers. The rollers are between "V" grooved bearing races, one being on the top carriage and the other on the base. Typically, bearing housings are constructed from aluminum while the rollers are constructed from steel.

Although roller slides are not self-cleaning, they are suitable for environments with low levels of airborne contaminants such as dirt and dust. As one of the more expensive types of linear slides, roller slides are capable of providing linear motion on more than one axis through stackable slides and double carriages. Roller slides offers line contact versus point contact as with ball bearings, creating a broader contact surface due to the consistency of contact between the carriage and the base and resulting in less erosion.

Plain Bearing

Plain bearings are very similar in design to rolling-element bearings, except they slide without the use of ball bearings. If they are cylindrical in shape, they are often called bushings. Bushings can be metal or plastic, or even air.

- Plain bearings can run on hardened steel or stainless steel shafting (raceways), or can be run on hard-anodized aluminum or soft steel or aluminum. For plastic bushings, the specific type of polymer/fluoro-polymer will determine what hardness is allowed.
- Plain bearings are less rigid than rolling-element bearings.
- Plain bearings handle contamination well and often do not need seals/scrapers.
- Plain bearings generally handle a wider temperature range than rolling-element bearings.
- Plain bearings (plastic versions) do not require oil or lubrication (often it can be used to increase performance characteristics).

Dovetail Slides

Dovetail slides, or dovetail way slides are typically constructed from cast iron, but can also be constructed from hard-coat aluminum, acetal or stainless steel. Like any bearing, a dovetail slide is composed of a stationary linear base and a moving carriage. a Dovetail carriage has a v-shaped, or dovetail-shaped protruding channel which locks into the linear base's correspondingly shaped groove. Once the dovetail carriage is fitted into its base's channel, the carriage is locked into the channel's linear axis and allows free linear movement. When a platform is attached to the carriage of a dovetail slide, a dovetail table is created, offering extended load carrying capabilities.

Dovetail slides are advantageous when it comes to load capacity, affordability and durability. Capable of long travel, dovetail slides are more resistant to shock than other bearings, and they are mostly immune to chemical, dust and dirt contamination. Dovetail slides can be motorized, mechanical or electromechanical. Electric dovetail slides are driven by a number of different devices, such as ball screws, belts and cables, which are powered by functional motors such as stepper motors, linear motors and handwheels. Dovetail slides are direct contact systems, making them fitting for heavy load applications including CNC machines, shuttle devices, special machines and work holding devices. Mainly used in the manufacturing and laboratory science industries, dovetail slides are ideal for high-precision applications.

Compound Slides

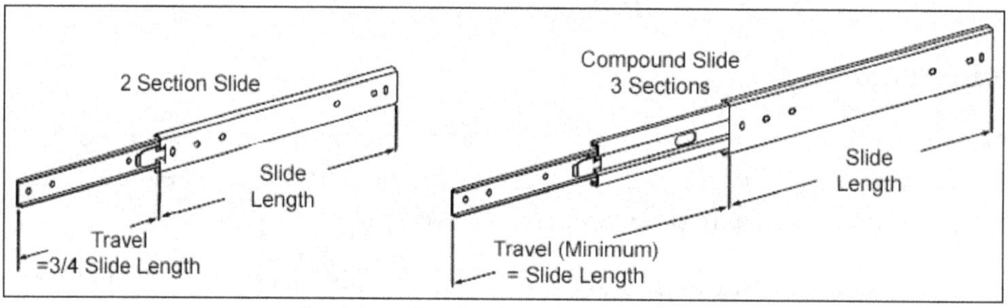

2-Section and 3-Section Compound Slide.

Slides can be constructed with two sections or multiple sections. A slide with two sections can only extend approximately 3/4 of the total compressed slide length. A compound slide typically has three sections: Fixed, floating intermediate member, and the section attached to the equipment. A compound slide can extend at least as far as the compressed slide length and typically a bit more. In the case of rack slides, this allows the equipment to extend completely out of the rack allowing access for service or connection of cables and such to the back of the equipment.

Rack Slides

Rack slides are specifically intended for mounting equipment into 19-inch racks or 23-inch racks. These can be friction bearing, ball bearing, or roller bearing. They are sized to fit into racks with mounting flanges on the ends to mate to the mounting holes in racks. In some cases, one mounting flange is formed into the rack slide with an adapter bracket attached to the other end to accommodate different depths of the rack. The outer fixed member is attached to the rack and the inner moving member is generally screwed to the side of the mounted equipment. Rack slides are typically compound or 3-part slides allowing full extension of the mounted equipment and generally include provision for sliding the inner member completely free to allow removal of the equipment from the rack. They can also include stops to prevent accidentally pulling the equipment out of the rack without releasing the stop mechanism.

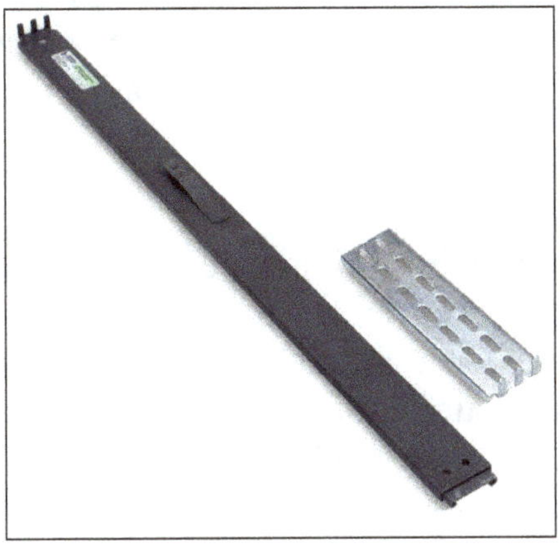

Friction Bearing Rack Slide.

There can be proprietary configurations which, for example, may clip to the equipment without the use of screws or can be clipped into an appropriately designed rack. But the basic geometry is the same regardless of how they are mounted.

Spherical Bearing

Spherical plain bearing.

Spherical bearing with rolling elements at the contact surface.

A spherical plain bearing is a bearing that permits angular rotation about a central point in two orthogonal directions (usually within a specified angular limit based on the bearing geometry). Typically these bearings support a rotating shaft in the bore of the inner ring that must move not only rotationally, but also at an angle.

Self-aligning spherical bearings were first used by James Nasmyth around 1840 to support line shaft bearings in mills and machine shops. For long shafts it was impossible to accurately align bearings, even if the shaft was perfectly straight. Nasmyth used brass bearing shells between hemispherical brass cups to align the bearings to self-align.

Construction

Spherical bearings can be of a hydrostatic or mechanical construction. A spherical bearing by itself consists of an outer ring and an inner ring and a locking feature that makes the inner ring captive within the outer ring in the axial direction only. The outer surface of the inner ring and the inner surface of the outer ring are spherical (or more correctly, toroidal) and are collectively considered the raceway and they slide against each other, either with a lubricant, a maintenance-free (typically polytetrafluoroethylene or PTFE) based liner, or they incorporate a rolling element such as a race of ball-bearings, allowing lower friction.

Application

Spherical bearings are used in countless applications, wherever rotational motion must be allowed to change the alignment of its rotation axis. A good example is the drive axle bearings of a vehicle control arm (or A-arm) suspension. The mechanics of the suspension allow the axle to move up and down (and the wheel to turn in order to steer the vehicle), and the axle bearings must allow the rotational axis of the axle to change without binding. While in practice, spherical bearings are rarely used here, it is a simple concept that illustrates a possible application of a spherical bearing. In fact, spherical bearings are used in smaller sub-components of this type of suspension, for example certain types of constant-velocity joints.

Spherical bearings are used in car suspensions, engines, driveshafts, heavy machinery, sewing machines, robotics and many other applications.

Cannon Bearing

A cannon bearing or cannon box bearing is an arrangement of bearings on a shaft, usually an axle, where two bearings are mounted in an enclosed tube.

The function of the cannon box is to preserve the alignment of the two bearings, even if the overall tube is allowed to move. The two bearings will retain their same relative position. The bearing tube can be attached to the vehicle frame through either a pivot or springs.

The name 'cannon box' derives from the appearance of the hollow tube. Also from the boring machines used to machine the accurately aligned bearing seats, the same machines developed for the boring of cannon and also used for machining the cylinders of steam engines.

Cannon box bearings are still found today, although much of the need for them was removed by the development of self-aligning ball bearings. These allow the inner race of a bearing to move independently of its outer, so that each bearing can align to both shaft and housing simultaneously, even if the housing moves around.

Railway Use

The 'cannon box' bearing was introduced by George Stephenson on his locomotive Locomotion of 1825 for the Stockton and Darlington Railway. One of the axles was carried in a 'cannon box' bearing that was pivoted centrally and could tilt from side to side. Unlike Stephenson's earlier locomotives and their steam springs, this new suspension had no springing but was a compensated suspension: As one side was raised, the bearing tube pivoted and the other side was pushed downwards. Although not giving a stable ride for the locomotive, it did allow the wheels to follow uneven track. Rails of this period were short pieces of cast iron rather than lengths of rolled steel and had frequent, often misaligned, joints.

A side-effect of the cannon box is that the centre of the shaft is now enclosed and inaccessible. Previously Stephenson had coupled the driving axles of his locomotives with a central chain. The need for an alternate system spurred the development of the coupling rods, where the drive was connected on the outside face of the wheels.

The enclosure of the shaft and bearing also had advantages for enclosing the bearings to exclude dirt and to contain lubricating oil.

Francis Webb's form of radial axle design for the LNWR used a cannon box. A radial axle is a form of single axle pony truck, where instead of a pivoted frame, the two axle bearings can slide sideways in a circular track. Webb's design placed them into a curved cannon box. David Joy, designer of the eponymous valvegear, described encountering these axleboxes on Webb's Precedent class. The earlier Adams design for a similar axle had relied on the axle and thrust-faces within the axle bearings to keep the hornblocks in position.

In the 1920s, some electric locomotives used rigid frames with a leading and trailing truck, where a cannon bearing was a convenient inside-framed bearing for the truck. The Swiss 'Java bogie' design, developed by Jakob Buchli, even used this for a driving axle, with the drive provided by a one-sided Buchli drive outside the wheels.

Roller bearings, and their enclosing cannon boxes, were used for the British Railways Standard classes, on bogie axles and also on the driving axles of some of the larger classes. This use was simplified, as the Standard classes had just two outside cylinders. Railway use of such bearings continues today. The Peppercorn Class A1 Tornado replica uses roller bearings rather than the original plain bearings and these use cannon tube bearings for both the driving and bogie axles, although not for the central cylinder's crank axle.

The rigidity of the cannon box also allows the faces of the axleboxes which move up and down in the hornguides to only need a single horn flange. Conventional axleboxes have a pair of flanges with the guide between them, so that they can move up and down for suspension without any axial movement. If they are both held at the correct spacing by the cannon tube, only one flange is needed. The thinner axlebox, without the external flange, can allow the hornguides and frames to be placed closer to the inner face of the wheel, giving better support for a heavily-loaded axle.

Traction Engines

Traction engines began as agricultural vehicles and ignored springing or suspension on their main wheels. As light steam tractors developed from 1896 for road haulage though, locomotives with solid rubber tyres and suspension became favoured. The driving wheels of a traction engine are carried on a single axle, running closely behind the rear plate of the boiler firebox and carried in bearings fitted to the hornplates on either side of the boiler. Where springs were used, the cannon box bearing was favoured for maintaining the bearing alignment, with a further advantage of enclosing the axle away from the coal and ash from the firebox. Designs such as the Tasker Little Giant used another compensated suspension, where the tube was suspended from a single transverse leaf spring.

Flexure Bearing

A flexure bearing is a category of flexure which is engineered to be compliant in one or more angular degrees of freedom. Flexure bearings are often part of compliant mechanisms. Flexure bearings serve much of the same function as conventional bearings or hinges in applications which require angular compliance. However, flexures require no lubrication and exhibit very low or no friction.

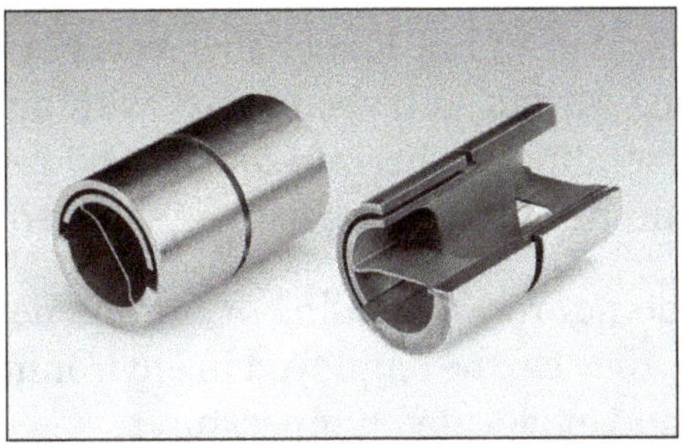

Flexure pivot: Utilized in place of bearings for their frictionless properties in precision alignment mechanisms and scientific instruments.

Many flexure bearings are made of a single part: two rigid structures joined by a thin "hinge" area. A hinged door can be created by implementing a flexible element between a door and the door frame, such that the flexible element bends allowing the door to pivot open.

Flexure bearings have the advantage over most other bearings that they are simple and thus inexpensive. They are also often compact, lightweight, have very low friction, and are easier to repair without specialized equipment. Flexure bearings have the disadvantages that the range of motion is limited, and often very limited for bearings that support high loads.

A flexure bearing relies on the bearing element being made of a material which can be repeatedly flexed without disintegrating. However, most materials fall apart if flexed a lot. For example, most metals will fatigue with repeated flexing, and will eventually snap. Thus, one part of flexure bearing design is avoiding fatigue.

Flexure bearings can give very low friction and also give very predictable friction. Many other bearings rely on sliding or rolling motions (rolling-element bearings), which are necessarily uneven because the bearing surfaces are never perfectly flat. A flexure bearing operates by bending of materials, which causes motion at microscopic level, so friction is very uniform. For this reason, flexure bearings are often used in sensitive precision measuring equipment.

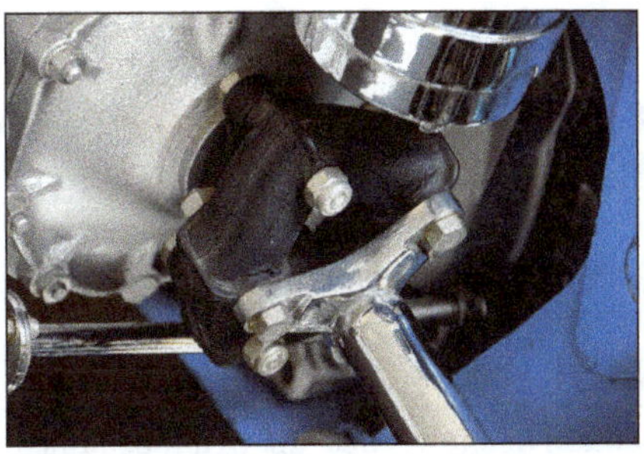

A Giubo driveshaft coupling, another type of flexure bearing, on the right hand rear driveshaft of a formula 2 race car. This coupling has two compliant degrees of freedom to allow rotation of the shaft with some misalignment.

Flexure bearings are not limited to low loads, however. For example, the drive shafts of some sports cars replace cardan universal joints with an equivalent joint called a rag joint which works by bending rubberized fabric. The resulting joint is lighter yet is capable of carrying hundreds of kilowatts, with adequate durability for a sports car.

Because flexure bearings do not rely on sliding or rolling motions, they do not require lubrication. Consequently, they can be employed in environments hostile to lubricants: underwater, in a vacuum and at elevated temperatures.

Jewel Bearing

Ruby jewel bearings used for a balance wheel in a mechanical watch movement.

A jewel bearing is a plain bearing in which a metal spindle turns in a jewel-lined pivot hole. The hole is typically shaped like a torus and is slightly larger than the shaft diameter. The jewel material is usually synthetic sapphire or ruby (corundum). Jewel bearings are used in precision instruments where low friction, long life, and dimensional accuracy are important. Their largest use is in mechanical watches.

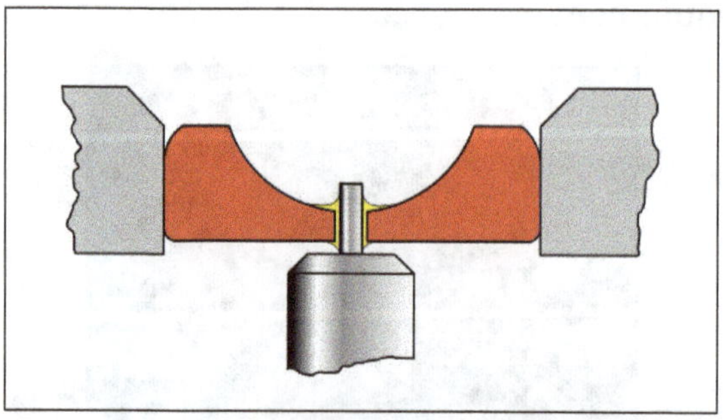

Cross-section of a jewel bearing in a mechanical watch.

This type of donut-shaped bearing (red) is called a hole jewel, used for most of the ordinary wheels in the gear train. It is usually made of synthetic sapphire or ruby, press-fit into a hole in the movement's supporting plate (grey). The cup-shaped depression in the top of the jewel is the oil cup; its purpose is to hold the lubricating oil (yellow) in contact with the bearing shaft by capillary action.

Characteristics

The advantages of jewel bearings include high accuracy, very small size and weight, low and predictable friction, good temperature stability, and the ability to operate without lubrication and in corrosive environments. They are known for their low kinetic friction and highly consistent static friction. The static coefficient of friction of brass-on-steel is 0.35, while that of sapphire-on-steel is 0.10–0.15. Sapphire surfaces are very hard and durable, with Mohs hardness of 9 and Knoop hardness of 1800, and can maintain smoothness over decades of use, thus reducing friction variability. Disadvantages include brittleness and fragility, limited availability/applicability in medium and large bearing sizes and capacities, and friction variations if the load is not axial.

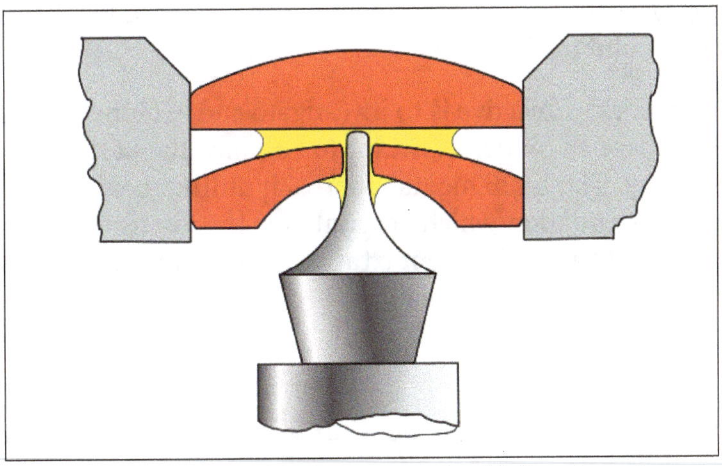

In wheels where friction is critical, a capstone is added on the end to prevent the shoulder of the shaft from bearing against the face of the jewel.

The predominant use of jewel bearings is in mechanical watches, where their low and predictable friction improves watch accuracy as well as improving bearing life. Manufacturers traditionally listed the number of jewels prominently on the watch face or back, as an advertising point. A typical fully jeweled time-only watch has 17 jewels: Two cap jewels, two pivot jewels and an impulse jewel for the balance wheel, two pivot jewels and two pallet jewels for the pallet fork, and two pivot jewels each for the escape, fourth, third, and center wheels. In modern quartz watches, the timekeeper is a quartz crystal in an electronic circuit, so accuracy of timekeeping is not dependent on low friction of the mechanical parts and jewels are not often used.

The other major use of jeweled bearings is in sensitive measuring instruments. They are typically used for delicate linkages that must carry very small forces, in instruments such as galvanometers, compasses, gyroscopes, gimbals, dial indicators, dial calipers, and turbine flow meters. Bearing bores are typically smaller than 1 mm and support loads weighing less than 1 gram, although they are made as large as 10 mm and may support loads up to about 500 g.

Pressure Distribution

Most of the modern lubricants consist of polymeric additives that can influence the base lubricants to slight shear thinning and mostly viscoelastic. This further classifies the fluid as non-Newtonian, which exhibits non-linearity in the relationship between shear stresses and shear rates. Many researchers have studied to determine the possible effects of such non-Newtonian lubricants may have a significance on the performance of bearings. The analysis show that such fluid played an important role on lubricating of moving parts. Lin studied the effect of lubricant rheology on hydrodynamic journal bearings, and approached the non-Newtonian behavior of the lubricating oil based on modern continuum point of view.

Hence, the present work's objective is to investigate and explore the effect of additives blended within the base as binder lubricant to enhance the performance of hydrodynamic journal bearing. The fundamental approach utilized here incorporates a modified form of Reynolds equation which accounts for the viscosity and volume fraction effects of base lubricants and additive mixture. Comparisons between Newtonian and non-Newtonian fluid effects are analyzed.

Analysis

Modifying the viscosity of base lubricants with addition of volume fraction v_2 of additive materials.

$$\mu = \mu_1 v_1 + \mu_2 v_2$$

Where μ_1 and μ_2 are viscosities of the binder lubricant and the additive respectively, v_1 and v_2 are corresponding volume fraction.

Where,

$$v_1 + v_2 = 1$$

$$\mu = (1 - v_2)\mu_1 + \mu_2 v_2$$

The dimensionless form of equation $\mu = (1 - v_2)\mu_1 + \mu_2 v_2$ can be acquired as,

$$\bar{\mu} = 1 + v_2(\eta - 1)$$

Where, $\bar{\mu}$ is the dimensionless viscosity and η is viscosity ratio of additive to the base binder.

Finally, using the non-dimensional mean film pressure and incorprating the viscosity viration due to volume fraction:

$$P = -12 \times \lambda^2 \times \varepsilon \times \bar{\mu} \times h^{-(n+2)} \times \left(Z^2 - \frac{1}{4} \right) \times \sin\theta$$

Results

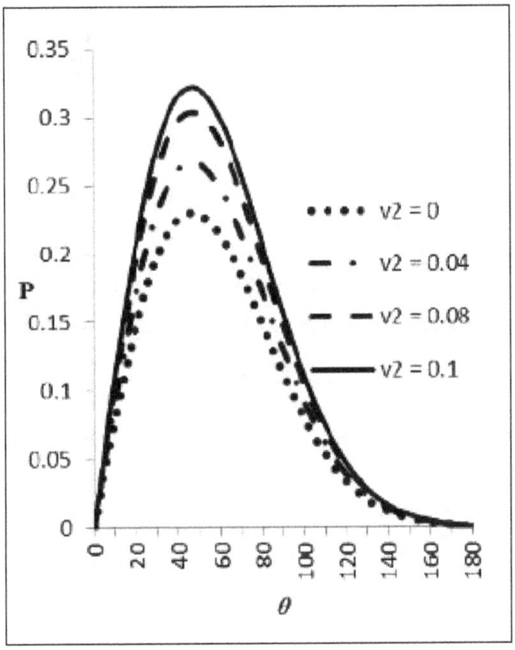

P vs θ under various volume fraction.

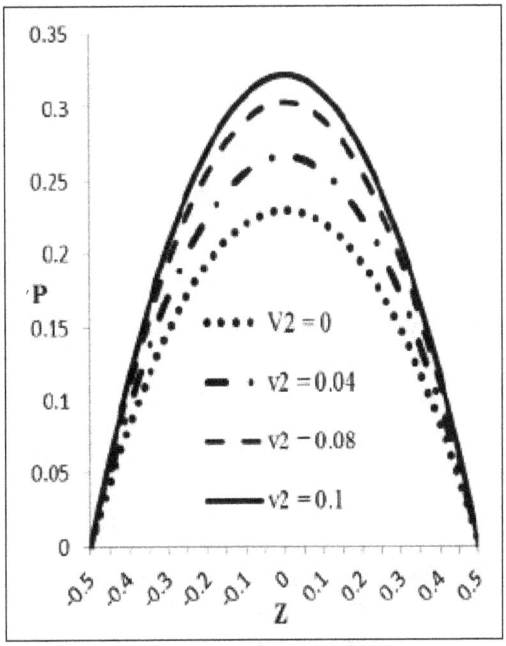

P vs Z under various volume fraction.

A pair of volume fraction of additives blended with the base lubricants have been used with different viscosity ratio. Dimensionless pressure distribution over the lubricant film has been identified. Figures show the relationship between dimensionless pressure (P) versus circumferential coordinate (θ) and dimensionless coordinate in the z-direction (Z) respectively. The plots generated under different volume fraction of additives. The lubricant fluid treated as Newtonian in both figures. The pressure generated in the fluid film is low without addition of additives ($v_2 = 0$). However, at each increase of volume fraction of additives ($v_2 = 0.04, 0.08, 0.1$) blended with the lubricants, the pressure lubrication increases in the fluid film. This increase of pressure in the fluid film in the journal bearing prevents contact between relatively moving parts of machineries. In return, it also improves the load carrying capacity of journal bearing.

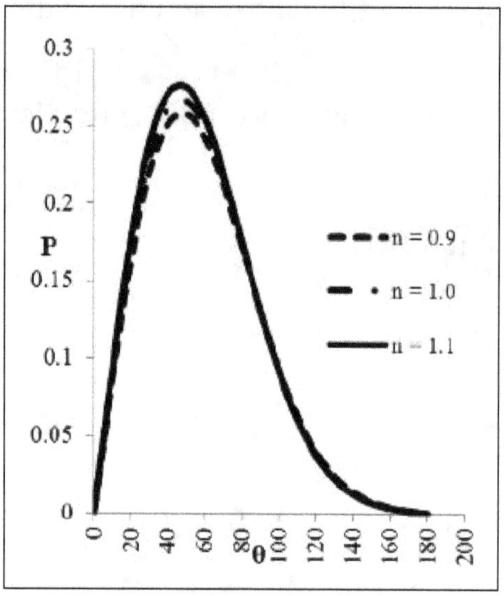

P vs θ under various flow behavior.

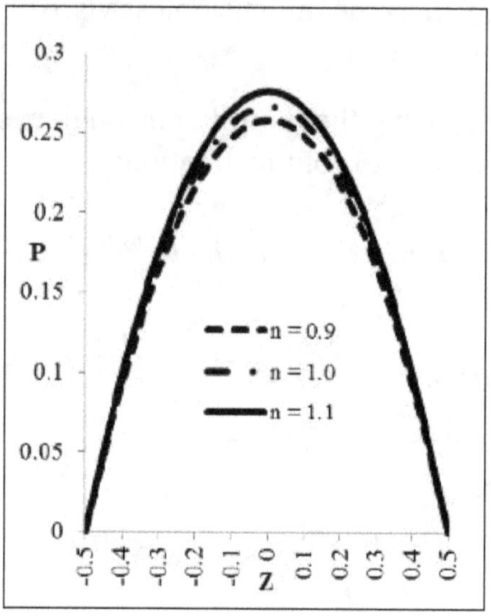

P vs Z under various flow behavior index.

On the other hand, figures are generated under various power law indexes, whereby, the fluid lubricant treated as non-Newtonian. Both show the relationship between dimensionless pressure (P) versus the circumferential coordinate (θ) and dimensionless coordinate in the z-direction (Z) respectively. The results have the same trend as in figures above with slight deviation from the Newtonian assumptions. It is concluded that, the pressure distribution in the fluid film is directly proportional to the portion of blended additives.

Thrust Bearing

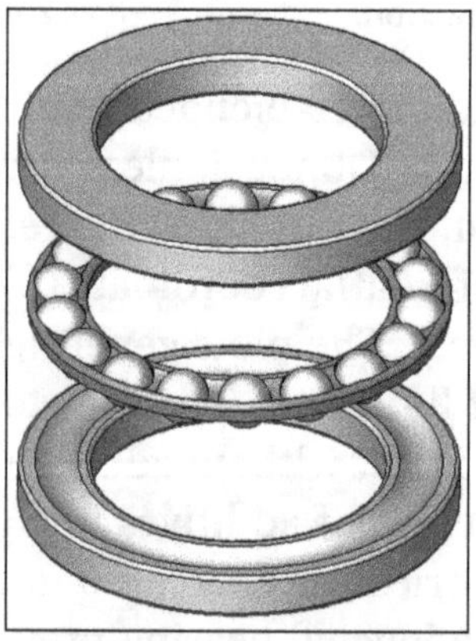

A thrust ball bearing.

A thrust bearing is a particular type of rotary bearing. Like other bearings they permit rotation between parts, but they are designed to support a predominantly axial load.

Thrust bearings come in several varieties:

- Thrust ball bearings, composed of bearing balls supported in a ring, can be used in low thrust applications where there is little axial load.
- Cylindrical thrust roller bearings consist of small cylindrical rollers arranged flat with their axes pointing to the axis of the bearing. They give very good carrying capacity and are cheap, but tend to wear due to the differences in radial speed and friction which is higher than with ball bearings.
- Tapered roller thrust bearings consist of small tapered rollers arranged so that their axes all converge at a point on the axis of the bearing. The length of the roller and the diameter of the wide and the narrow ends and the angle of rollers need to be carefully calculated to provide the correct taper so that each end of the roller rolls smoothly on the bearing face without skidding. These are the type most commonly used in automotive applications (to support the wheels of a motor car for example), where they are used in pairs to accommodate axial thrust in either direction, as well as radial loads. They can support greater thrust loads than the ball type due to the larger contact area, but are more expensive to manufacture.

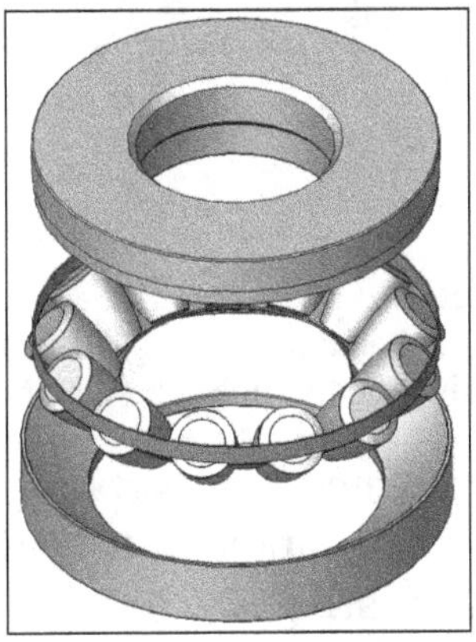

A spherical roller thrust bearing.

- Spherical roller thrust bearings use asymmetrical rollers of spherical shape, rolling inside a house washer with a raceway with spherical inner shape. They can accommodate combined radial and axial loads and also accommodate misalignment of the shafts. They are often used together with radial spherical roller bearings. Spherical roller thrust bearings offer the highest load rating density of all thrust bearings.
- Fluid bearings, where the axial thrust is supported on a thin layer of pressurized liquid—these give low drag.
- Magnetic bearings, where the axial thrust is supported on a magnetic field. This is used where very high speeds or very low drag is needed, for example the Zippe-type centrifuge.

Thrust bearings are commonly used in automotive, marine, and aerospace applications. They are also used in the main and tail rotor blade grips of RC (radio controlled) helicopters. Thrust bearings are used in cars because the forward gears in modern car gearboxes use helical gears which, while aiding in smoothness and noise reduction, cause axial forces that need to be dealt with. Thrust bearings are also used with radio antenna masts to reduce the load on an antenna rotator.

One specific thrust bearing in an automobile is the clutch "throw out" bearing, sometimes called the clutch release bearing.

Fluid Film Thrust Bearings

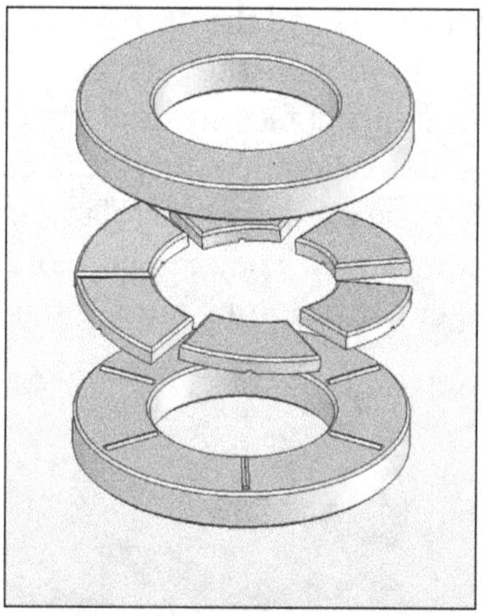

Exploded view of a Michell type thrust bearing. Each sector shaped pad can pivot on the ridges on the lower plate.

Fluid-film thrust bearings were invented by Albert Kingsbury, who discovered the principle in the course of bearing and lubrication investigations commencing in 1888 while a student. His first experimental bearing was tested in 1904. He filed for a patent in 1907, and it was granted in 1910. The first Kingsbury bearing in hydroelectric service, one of its major applications, was installed at the Holtwood Generating Station in 1912. It remains in full use today.

Thrust bearings were independently invented by Australian engineer George Michell (pronounced Mitchell) who patented his invention in 1905.

Fluid thrust bearings contain a number of sector-shaped pads, arranged in a circle around the shaft, and which are free to pivot. These create wedge-shaped regions of oil inside the bearing between the pads and a rotating disk, which support the applied thrust and eliminate metal-on-metal contact.

Kingsbury and Michell's invention was notably applied to the thrust block in ships. The small size (one-tenth the size of old bearing designs), low friction and long life of Kingsbury and Michell's invention made possible the development of more powerful engines and propellers. They were used extensively in ships built during World War I, and have become the standard bearing used on turbine shafts in ships and power plants worldwide.

Tilting Pad Thrust Bearings

Tilting pad thrust bearings are designed to transfer high axial loads from rotating shafts with minimum power loss, while simplifying installation and maintenance. The shaft diameters for which the bearings are designed range from 20 mm to more than 1,000 mm. The maximum loads for the various bearing types range from 0.5 to 500 tons. Bearings of larger size and load capacity are considered nonstandard but can be made to special order.

Each bearing consists of a series of pads supported in a carrier ring; each pad is free to tilt so that it creates a self-sustaining hydrodynamic film. The carrier ring may be in one piece or in halves with various location arrangements.

Multiple Options

Two options exist for lubrication. The first is to fully flood the bearing housing. The second, which is more suitable for higher speed applications, directs oil to the thrust face. This oil is then allowed to drain freely from the bearing housing.

Similarly, two geometric options exist. The first option does not use equalizing or leveling links. This option is used in many gear units and other shaft systems where perpendicularity between shaft centerline and bearing faces is assured.

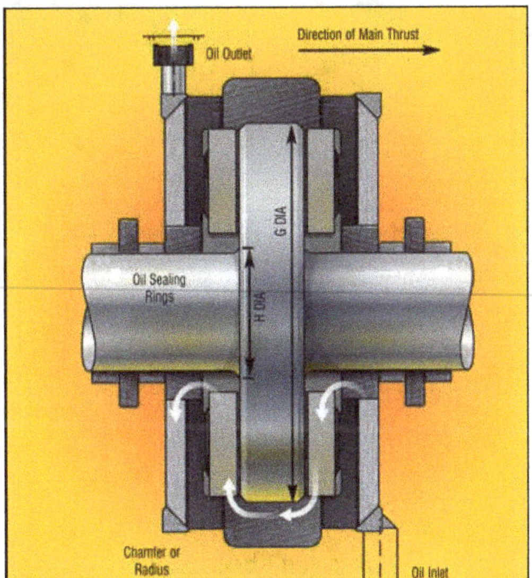

Flooded Lubrication: Typical Double-thrust Arrangement.

Bearings for both flooded and directed lubrication are intended for machines where an equalized thrust bearing is specified by API requirements, or where the bearing may be required for other reasons.

Flooded vs. Directed Lubrication

The conventional method of lubricating tilting pad thrust bearings is to flood the housing with oil, using an orifice on the outlet to regulate the flow and maintain pressure. A housing pressure of 0.7 to 1.0 bar (10.1 to 14.5 PSI) is typical, and to minimize leakage, seal rings are required where the shaft passes through the housing.

Although flooded lubrication is simple, it results in high parasitic power loss due to turbulence at high speed. Where mean sliding speeds in excess of 50 meters per second (m/s) are expected, these losses may be largely eliminated by employing the system of directed lubrication. As well as reducing power loss by typically 50 percent, directed lubrication reduces the bearing temperature, and in most cases, oil flow.

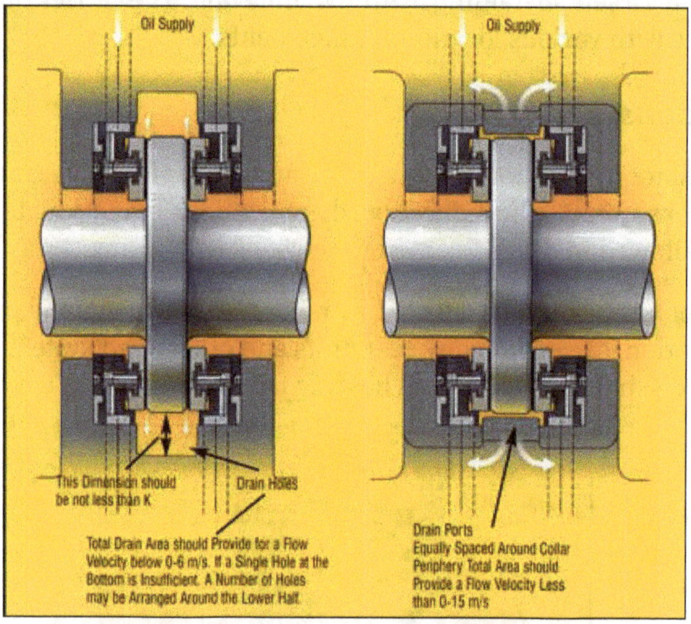

Directed Lubrication: Typical Double-thrust Arrangements Designed to Prevent Bulk Oil from Contacting the Collar.

It should be noted that:

- Directed and flooded bearings have the same basic sizes and use identical thrust pads.
- Preferred oil supply pressure for directed lubrication is 1.4 bar (20.3 PSI).
- Oil velocity in the supply passages should not exceed three meters per second (m/s) to ensure full pressure at the bearing.
- The bearing housing must be kept free of bulk oil through an ample drain area around the collar periphery.
- No seal rings are required on the shaft.

Manufacturers offer a variety of pad materials. Some polymeric materials are capable of operating at temperatures up to 120 °C (248 °F) higher than conventional white metal or babbitt. Also, pad pivot position can have an effect on thrust pad temperature.

All pads can be supplied with offset pivots, but center-pivoted pads are preferred for bidirectional running, foolproof assembly and minimum stocks. At moderate speeds, the pivot position does not affect load capacity; however, where mean sliding speeds exceed 70 m/s, offset pivots can reduce bearing surface temperatures and thus increase load capacity under running conditions.

Thrust bearings can be fitted with temperature sensors, proximity probes and load cells.

In hydraulic thrust metering systems, a hydraulic piston is located behind each thrust pad and is connected to a high-pressure oil supply. The pressure in the system then gives a measure of the applied thrust load. Figure shows a typical installation of this system complete with control panel, which incorporates the high-pressure oil pump and system pressure gauge calibrated to read thrust load.

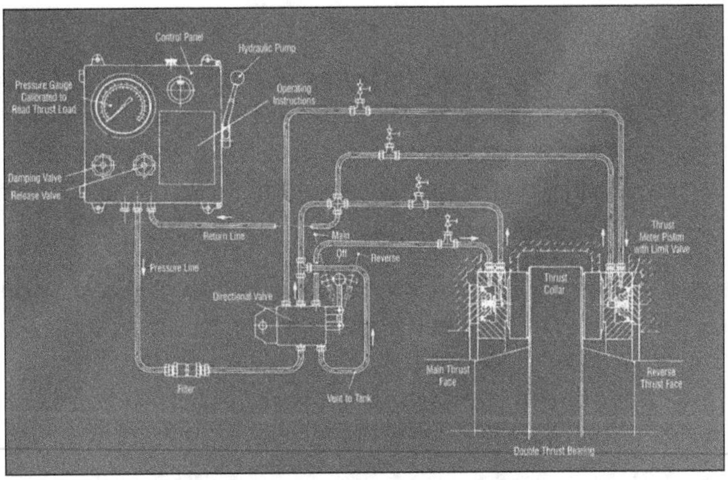

Hydraulic Thrust Metering Arrangement.

For systems incorporating load cells or hydraulic pistons, it is typically necessary to increase the overall axial thickness of the thrust ring.

Finally, thrust bearings incorporate hydraulic jacking provisions. These provisions ensure that an appropriate oil film exists between thrust runner and bearing pads while operating at low speeds.

At startup, the load-carrying capacity of tilting pad thrust bearings is restricted to approximately 60 percent of the maximum permissible operating load. If the startup load on a bearing exceeds this figure and a larger bearing is not an option, the manufacturer can supply thrust bearings fitted with a hydrostatic jacking system to allow the bearing to operate with heavy loads at low speeds. This system introduces oil at high pressure (typically 100 to 150 bar (1,450 to 2,175 PSI) between the bearing surfaces to form a hydrostatic oil film. It should be noted that a similar approach is taken when making hydraulic jacking provisions for radial bearings.

The bearing housing requirements for the KingCole LEG bearing are similar to those of standard thrust bearings. Oil seals at the back of the carrier rings are not required because the inlet oil is confined to passages within the base ring assembly. Fresh oil enters the bearing through an annulus located at the bottom of the base ring. The discharge space should be large enough to minimize contact between the discharged oil and the rotating collar. The discharge oil outlet should be sized so that oil can flow freely from the bearing cavity.

The manufacturer recommends a tangential discharge opening, equal in diameter to 80 percent of the recommended collar thickness. If possible, the discharge outlet should be located in the bottom of the bearing housing. Alternatively, it should be located tangential to the collar rotation. The bearing pads and carrier ring are constructed so that cool undiluted inlet oil flows from the leading edge groove in the bearing pad directly into the oil film. The cool oil in the oil film wedge insulates the white metal face from the hot oil carryover that adheres to the rotating collar.

In contrast to the LEG bearing, the oil for spray-fed bearings is injected between the bearing surfaces, not directly on them. This can result in uneven bearing lubrication and the need to supply nonpractical high pressure to achieve true effective scouring of the hot oil carryover adhering to the thrust collar. There is also a possibility for the small jet holes to clog with foreign material.

Friction power loss is claimed to be lower than both flooded and spray-fed bearings due to the reduced oil flow. The flow of cool oil over the leading edge lowers pad surface temperatures and increases the KingCole's capacity.

The resulting performance improvements are shown in figures:

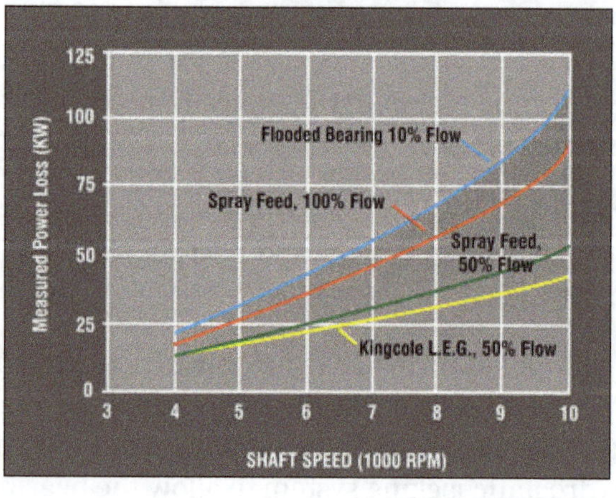

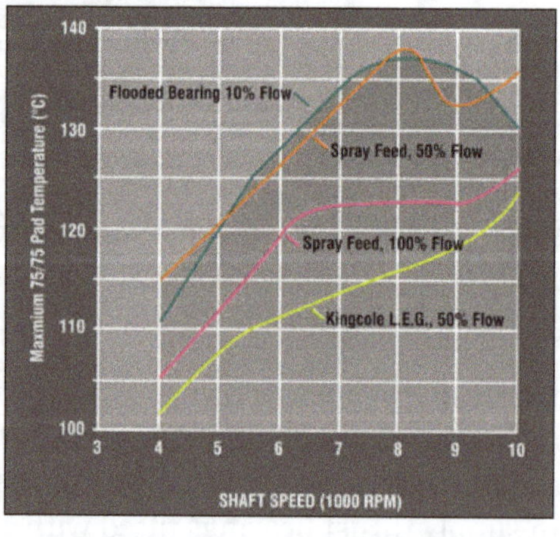

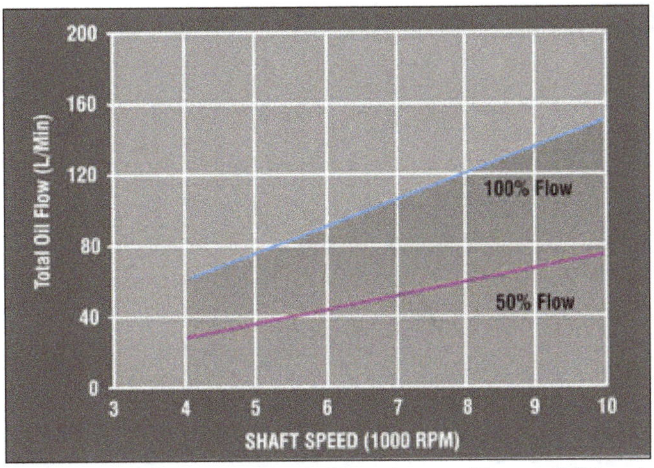

LEG Bearings vs. Standard Flooded Bearings and Spray-fed Bearings.

Assuming an oil inlet temperature of 50 °C (122.4 °F), it is possible to estimate the white metal temperature of KingCole leading edge bearings from figure. These temperatures are a function of surface speed and contact pressure.

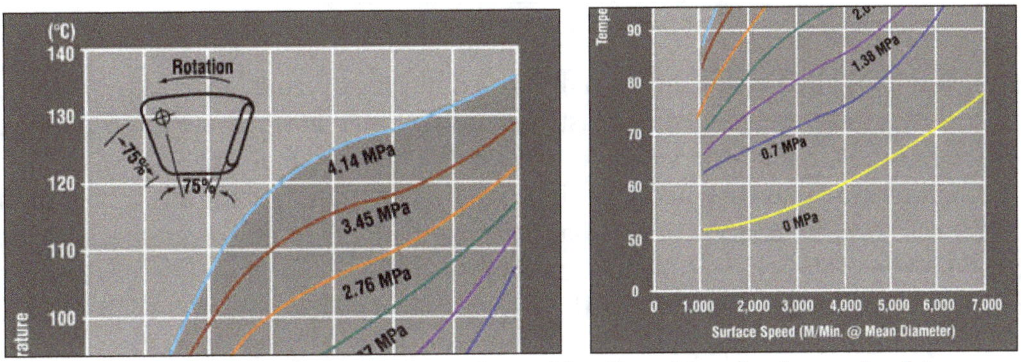

LEG White Metal Temperatures at 75/75 Position (6- and 8-pad series, steel pads).

Bearing Selection

Thrust load, shaft RPM, oil viscosity and shaft diameter through the bearing determine the bearing size to be selected. Leading edge bearings are sized for normal load and speed when transient load and speed are within 20 percent of normal conditions.

All curves are based on an oil viscosity of ISO VG32, with an inlet oil temperature of 50 °C (122.4°F). The manufacturer recommends ISO VG32 oil viscosity for moderate- through high-speed applications.

Thrust Block

A thrust block, also known as a thrust box, is a specialised form of thrust bearing used in ships, to resist the thrust of the propeller shaft and transmit it to the hull.

Early Thrust Boxes

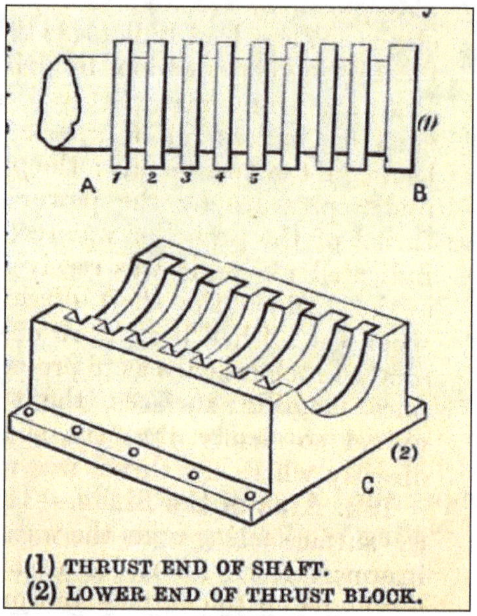

Multi-collar thrust box, with shaft.

Early screw-propelled steamships used a thrust block or thrust box composed of perhaps a dozen lower-rated plain thrust journal bearings stacked on the same shaft. These were problematic in service: they were bulky, difficult to dismantle, wasted power through friction and they had a tendency to overheat. The thrust box was built of a box-like cast iron housing with a radial bearing at each end and a number of collars formed on the shaft between them. This shaft was often a short section of removable shaft called the thrust shaft, linking the engine ahead to the propeller shaft astern. A series of iron horseshoe-shaped collars fitted over the small diameter of the shaft and bore against the forward face of the shaft's collars. Each horseshoe was faced with a low-friction pad of babbitt metal. Lubrication was by an oil bath in the box and a plentiful volume was important for cooling purposes too.

Although lignum vitae wood was used for the radial stave bearings in the stuffing box, cooled directly by seawater itself, this material wasn't capable of withstanding the force needed for the thrust blocks of any but the earliest screw vessels.

Each horseshoe was independently adjustable forwards and back, by either wedged gibs or a screwed adjustment. A particular problem with these thrust boxes was in adjusting them so that the force was shared equally between all the collars. Adjustment was often done on the basis of their operating temperature, gauged with the engineer's hand.

Improved Michell Thrust Blocks

Improved understanding of the theory of lubrication films (initially by Reynolds) allowed the development of much more efficient bearing surfaces. This allowed the replacement of multiple collars in a thrust box by a single thrust block.

Michell thrust blocks at the London Science Museum.

In 1905, Australian engineer George Michell obtained a patent for the thrust block. (Subsequently, American engineer Albert Kingsbury established that tests he conducted in 1898 predated Michell's work. In 1910, Kingsbury was awarded US patent No. 947242 for the fluid-film thrust bearing).

Michell bearings contain a number of sector-shaped pads, arranged in a circle around the shaft, and which are free to pivot. These create wedge-shaped films of oil between the pads and a rotating disk on the shaft. Each lubricant "wedge" can only be of a limited length (in the direction of travel, i.e. circumferential) so multiple pads are needed rather than a single ring. No lubrication pump is needed: The rotation of the shaft itself is sufficient.

The need for an efficient thrust block became even more important with the advent of steam turbines, with their higher propeller speeds. Despite this, there was some reluctance to adopt them in their homeland, until the discovery that World War I U-boats were using them. After this they were soon adopted widely. The large single pad illustrated is a model of one used in the battlecruiser HMS Hood, once the pride of the Royal Navy. Michell Bearings continue in production today under the same name, first as part of Rolls-Royce marine systems.

Rolling Elements

Contact Form

The rolling motion of machine elements under considerable loads can cause effects on viscosity of the lubricants and deformation in the contact zones. Hence, elastohydrodynamic lubrication is the predominant regime for this type of motion. This phenomena occurs widely with nonconforming contacts and it is crucial for highly-loaded contacts such as line contacts of a roller bearing, point contact for a ball bearing, also for point and line contact in the case of gear teeth. In this lubrication regime, factors such as temperature and shear rate have big influence on the behavior of the lubricant. In order to understand the tribological behavior of these machine elements, types of contacts has to be evaluated regarding loads and motion. Line and point contact are labeled as footprints contacts, line contacts occurs when 2 cylinders are in contacts, and as the load increases this lines can form a rectangular footprint. While the contact of a sphere and a flat surface causes the formation of circular points contacts. The contact of a ball and a raceway of a bearing under considerable loads, initiates the formation of elliptical footprints. The analysis of nonconforming contacts can be reduced to the case of a nonconforming surface and a plane surface. The following example shows the case of two cylinders with respective velocities u_a and u_b transformed to an equivalent cylinder having equivalent radius. It is also important to evaluate the relative sliding velocity as the difference $|u_a - u_b|$ and also the entraining velocity of interest known also as rolling velocity and defined as:

$$u = (u_a + u_b)/2$$

It is also useful to define the slide to roll ratio S as:

$$S = \frac{sli.velocity}{2rol.velocity} = \frac{u_a - u_b}{u_a + u_b}$$

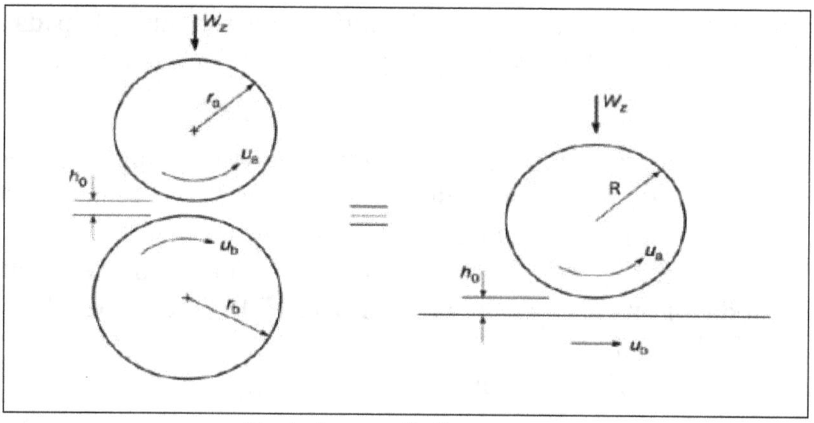

Equivalent 2 cylinders system.

Line Contact

Rigid Cylinder Contact

For a constant viscosity, the integration of Reynolds equation over the area of a cylinder moving in the x direction on a flat surface, yields that the film thickness has a parabolic function as:

$$h = h_0 + \frac{x^2}{2R} \text{ when } x \ll R$$

Where R is the equivalent radius (effective radius) and h_o is the minimum film thickness.

Considering the inlet and the outlet pressures to be ambient, the maximum pressure and the load capacity can be expressed as:

$$p_{max} = 2.15\eta_o u(\frac{R}{h_o^3})^{1/2}$$

$$W = 4.9\eta_o u\frac{R}{h}$$

For a variable viscosity and using Barus relation, it is useful to define a new variable called reduced pressure as,

$$p_r = \frac{1-e^{-\alpha p}}{\alpha}$$

Then the maximum pressure and corresponding film thickness can be simply defined as,

$$(p_r)_{max} = 2.15\eta_o u(\frac{R}{h_o^3})^{1/2}$$

$$h_o = 1.66(\alpha\eta_o u)^{2/3} R^{1/3}$$

The load capacity is found after integrating the area under the pressure curve.

Elastic Cylinder Contact

Elastic deformation occurs in nonconforming surfaces under any loads. The analyze of the lubricant pressure distribution shows that it is elliptical and near Hertzian pressure and that it is constant through the contact region. The evaluation of the film thickness for elastic contacts is done using elasticity and Reynolds equation as follows:

$$h = h_o + \frac{x^2}{2r} + w$$

Where x is the position, r is the equivalent radius of curvature for elliptical contact and w is combined deformation of the solids in contact. The minimum film thickness can be determined as,

$$\frac{h_o}{R} = 2.08(\frac{\eta_o u\alpha}{R})^{8/11}(\frac{E'R}{W})^{1/11} = 2.08(UG)^{8/11}(W')^{-1/11}$$

Where U is a speed parameter $\left(\frac{\eta_o u}{E'R}\right)$, G is a materials parameter $(\acute{a}E')$, W' is a load parameter ($\frac{W}{E'R}$) while E' is the composite modulus. Based on different investigations of elastohydrodynamic behavior, the following results for the minimum and the center film thickness have been made:

$$\frac{h_{min}}{R} = 1.714 U^{0.694} G^{0.568} (W')^{-0.128}$$

$$\frac{h_c}{R} = 2.922 U^{0.692} G^{0.47} (W')^{-0.166}$$

The Thickness of the film for the viscous case is greater than the isoviscous case and generally for a successful lubrication, the film thickness must be high compared to the surface roughness of the bearing.

Point Contact

The circular contact points occur as a result of two spheres contact while elliptical contact points manifest when a sphere comes into contact with a raceway. This high pressure spots induce the deformation at the contact interface and elevate the lubricant oil viscosity. Similarly tribilogical features for this type of contact can be solved using elasticity and Reynolds equation. The elastohydrodynamic behavior of point contacts for the case of high elastic modulus materials permitted the development of the relations below:

$$\frac{h_{min}}{R_x} = 3.63 U^{0.68} G^{0.49} W'^{-0.073} [1 - \exp(-0.68k)]$$

$$\frac{h_c}{R_x} = 2.69 U^{0.67} G^{0.53} W'^{-0.067} [1 - 0.61 \exp(-0.73k)]$$

For U is a speed parameter ($\frac{\eta_0 u}{E' R_x}$), G is a materials parameter $(\acute{a} E')$, W' is a load parameter ($\frac{W}{E' R_x}$) while E' is the composite modulus, K is the ellipticity parameter $(k = R_y / R_x)$ and R_y is the effective radii in y direction while Rx is the effective radii in the sliding direction x.

For materials with low elastic modulus the minimum film thickness and the center film thickness can be evaluated as following:

$$\frac{h_{min}}{R_x} = 7.43 U^{0.64} W'^{-0.21} [1 - 0.85 \exp(-0.31k)]$$

$$\frac{h_c}{R_x} = 7.32 U^{0.64} W'^{-0.22} [1 - 0.72 \exp(-0.28k)]$$

The following images shows examples of point and line contacts:

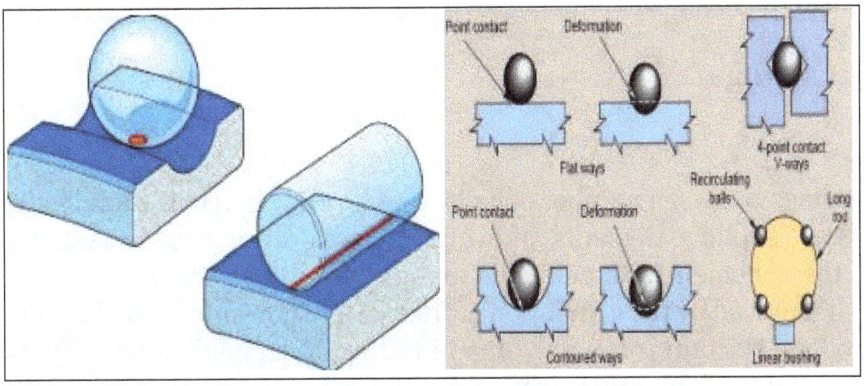

Contacts examples for sphere, cylinder and ball bearing.

Further studies showed that in presence of lubricant, the pressure distribution is almost Hertzian, hence the maximum pressure for a circular point contact can be evaluated as:

$$p_{\max} = \frac{3W}{2\pi a^2}$$

$$a = {}^{1/3}\sqrt{\frac{3RW}{4E'}}$$

Where W is the applied load, R is the equivalent radius, E' is the composite modulus and a is contact radius. Further formulas are set in figure.

Thermal Correction

The line and point contact theory presented before was in the case of isothermal conditions. However in most of the applications, the temperature increases at the contact interface what changes the liquid viscosity. Generally it is better to determine the temperature at the contact zone and to include it while setting the absolute viscosity of the lubricant. Moreover, engineering methods were developed to approximate temperature correction. In fact for point contact studies are not yet advanced while for line contact the following equation for thermal reduction factor is used:

$$T_r = 0{,}857 - 0.0234T_l + 0.000168(T_l)^2$$

$$T_l = \frac{\eta_0 \beta u^2}{\rho}$$

Where u is the entraining velocity, η_0 is the absolute viscosity at atmospheric pressure and surface temperature, ρ is the lubricant thermal conductivity and β is the viscosity-temperature coefficient. This thermal reduction factor is multiplied by the film thickness to estimate its true value. Different viscosity coefficients and values are to be measured at different pressures and temperatures to estimate true values of the film thickness, and the most used method for research is the rolling disk-type apparatus.

Rolling Bearings

Friction

Rolling bearings contain a number of balls or rollers that roll in the annular space. The contact between rolling elements and rings or races causes resistance to the relative motion of these elements. Actually, the motion of rolling elements over yielding surfaces is characterized by the generation of friction and unsymmetrical load distribution over the rollers. Friction is not only due to rolling or slip sliding but also to different factors. The total friction torque is the sum of the individual torques caused by mechanical and dynamic factors.

Differential Sliding

The rolling of rollers or balls along curved rings develops pure rolling, along 2 lines of an ellipse, and sliding because of difference in distances between contact points to the axis of rotation. This is called differential sliding that origins a friction torque what can be estimated as:

$$M_{ds} = \frac{\mu_s W D_0 Z}{4d_b}(1-\frac{d_b^2}{D_0^2}\cos^2\alpha)(r_0 B_0 - r_i B_i)$$

Where μ_s is the sliding coefficient of friction, W is the load,D_0 is the mean diameter, d_b is the ball diameter, Z is is the number of balls, α is the contact angle, r_i and r_0 are the radii of the outer and inner raceways while B_0 and B_i are factors based on geometric deformation.

Gyroscopic Spin

The contact angle α causes a friction torque that can be expressed as following:

$$M_{gr} = I\omega_b Z\sin\alpha = \frac{\pi}{60}d_b^2\frac{\gamma}{g}\omega_b Z\sin\alpha$$

Where I is the ball moment of inertia, ω_b is the angular velocity of the ball, γ is the density of the ball and g is the acceleration of gravity.

Elastic Hysteresis

There are energy losses due to elastic hysteresis of the material and the friction torque can be estimated as,

$$M_{hs} = 0.000125 D_0 d_b^{-2/3}\sum p_i^{4/3}$$

Where p_i is the load on the ith ball.

Raceway Torque

The weight of the inner ring causes contact with the roller in a focused point and the friction torque can be defined as following:

$$= \frac{{}_{s}G\ D}{4}(1-\frac{d_b}{}\cos^2\alpha)\sin[\alpha+\tan^{1}(\frac{d_b}{2}\sin\alpha)]$$

Where μ_c is the friction coefficient between the rolling element and the raceway housing, R_r is the raceway radius and G_c is the raceway housing mass.

Lubricant Shear

The presence of the lubricant induces an additional energy loss that is relative to the lubricant features, the friction torque generated by the shear of the lubricant for a cylindrical roller is:

$$M_{ls} = \frac{13Z\pi v^2\eta}{\sqrt{\lambda h_o}}$$

And for rolling ball is,

$$M_{ls} = \frac{12Z\pi v^2\eta}{(2\beta/3\lambda)\sqrt{\beta}}1n(\frac{4m^2\beta}{h_o})$$

$$\lambda = 0.5(\frac{1}{R_x}+\frac{1}{R_1})$$

$$\beta = 0.5(\frac{1}{R_y}+\frac{1}{R_2})$$

Where v is the longitudinal velocity of the contact point in rolling, m is half of the film thickness, R_x is the curvature radius of the roller, R_1 is the raceway radius of curvature While R_y R_2 are the radii perpendicular the motion direction.

There are also other factors contributing in the total torque friction such as geometric errors that causes deviation from the theoretical line of action, also friction can be generated by working medium environment and by temperature increase.

SKF Friction Model

SKF established models to estimate the exact torque of friction in rolling element bearings. The simplest model can be used when the bearing dynamic load P is 10 times smaller than the thrust load, then the total friction torque can be estimated as:

$$M_t = 0.5\mu P d$$

Where μ is the bearing coefficient of friction and d is the bearing bore diameter (F inner ring raceway diameter for needle roller bearings).

Generally SKF established a model to calculate friction on all rolling element types suggesting that the total torque is the sum of the rolling frictional moment, sliding friction moment, seals frictional moment and the friction moment caused by drag losses.

Rolling frictional moment M_r can be calculated using the following relation:

$$M_r = \Phi_{ish} \Phi_{rs} G_{rr} (n\eta)^{0.6}$$

Where Φ_{ish} is the inlet shear heating reduction factor, Φ_{rs} is the kinematic replenishment/starvation reduction factor and G_{rr} is a variable (depending on the bearing type the bearing mean diameter d_m[mm]= 0,5(d + D), the radial load F_r [N] the axial load F_a [N]) while n is the rotational speed [r/min] and η is the actual operating viscosity of the oil or the base oil of the grease [mm²/s].

The shear heating reduction can be calculated as,

$$\Phi_{ish} = \left[1 + 1.84\,10^{-9} (n\,d_m)^{1.28} \eta^{0.64}\right]^{-1}$$

$$\phi_{rs} = \frac{1}{e^{\left[k_{rs} n\eta (d+D) \sqrt{\frac{k_z}{2(D-d)}}\right]}}$$

Where K_{rs} is a replenishment/starvation constant (3 10^{-8} for low level oil bath and oil jet lubrication, 6 10^{-8} for grease and oil-air lubrication), K_z is a geometric constant related to the bearing type, d is the bearing bore diameter [mm] and D is the bearing outside diameter [mm].

Sliding frictional moment M_s can be calculated as following:

$$M_s = G_{sl}\, \mu_{sl}$$

Where G_{sl} is a variable depending on loads, type and diameter of the bearing and μsl is the sliding friction coefficient that is defined as,

$$\mu_{sl} = \Phi_{bl} \mu_{bl} + (1 - \Phi_{bl})\, \mu EHL$$

Where μ_{bl} is a coefficient depending on the additive package in the lubricant, generally ≈ 0.15, μEHL is the sliding friction coefficient in full-film conditions (0,02 for cylindrical roller bearings 0,002 for tapered roller bearings and 0,05 for lubrication with mineral oils 0,04 for lubrication with synthetic oils 0,1 for lubrication with transmission fluids) and Φ_{bl} is a weighting factor for the sliding friction coefficient defined as,

$$\phi_{bl} = \frac{1}{e^{2.6} (n\eta)^{1.4} 10^{-8} d_m}$$

The frictional moment of seals M_{seal} is calculated as,

$$M_{seal} = k_{s1} d_s^{\beta} + k_{s2}$$

Where K_{S1} is a constant depending on the seal type and the bearing type and size, d_s is the seal counter-face diameter [mm], β is an exponent depending on the seal type and the bearing type while K_{S2} is constant depending on the seal type and the bearing type and size.

The rotation of the bearing in oil bath causes drag losses and the friction torque M_{drag} for a roller bearing is,

$$M_{drag} = 0.4 V_m k_{roll} C_w B d_m^4 n^2 + 1.093 \times 10^{-7} n^2 d_m^3 \left(\frac{n d_m^2 f_t}{\eta}\right)^{-1.379} R_s$$

For a ball bearing,

$$M_{drag} = 0.4 V_m k_{ball} d_m^5 n^2 + 1.093 \times 10^{-7} n^2 d_m^3 \left(\frac{n d_m^2 f_t}{\eta}\right)^{-1.379} R_s$$

All the constants and the factors needed for the friction torques calculation are set in SKF catalogues for rolling elements bearings. The starting torque for this type of bearing can be estimated as the sum of the sliding frictional moment and seals frictional moment. Consequently the power loss generated from friction N_R and the temperature increase Δt are calculated as,

$$N_R = 1.05 10^{-4} M_t n$$

$$\Delta t = N_R / W_s$$

Where W_s is the cooling factor.

Deflection

In order to estimate the tribological features of rolling bearings, it is crucial to approximate their elastic deformation under applied load. Palmgren established the relations in the following figure to predict deflection of rolling bearings where P is the applied load, d is the rolling element diameter, l is the length of the contact and α is the bearing contact angle (the angle between the line drawn through rolling element and the radial plane of the bearing).

Table: Palmgren deflections.

	Radical deflection	Axial Deflection
Self-allign ball bearing	$\delta_r = \frac{6.858 \times 10^{-8}}{\cos\alpha} \left(\frac{P^2}{d}\right)^{\frac{1}{3}}$	$\delta_a = \frac{6.858 \times 10^{-8}}{\sin\alpha} \left(\frac{P^2}{d}\right)^{\frac{1}{3}}$

Deep-groovr and anugular bearing	$\delta_r = \frac{4.3\times10^{-8}}{\cos\alpha}\left(\frac{P^2}{d}\right)^{\frac{1}{3}}$	$\delta_a = \frac{4.3\times10^{-8}}{\sin\alpha}\left(\frac{P^2}{d}\right)^{\frac{1}{3}}$
Roller-bearing for point-line contact	$\delta_r = \frac{5.61\times10^{-9}}{\cos\alpha}\frac{\left(P^3\right)^{\frac{1}{4}}}{l^{\frac{1}{2}}}$	$\delta_a = \frac{5.61\times10^{-9}}{\sin\alpha}\frac{\left(P^3\right)^{\frac{1}{4}}}{l^{\frac{1}{2}}}$
Roller-bearing for line contact	$\delta_r = \frac{3.0\times10^{-10}}{\cos\alpha}\frac{P^{0.9}}{l^{0.8}}$	$\delta_a = \frac{3.0\times10^{-10}}{\sin\alpha}\frac{P^{0.9}}{l^{0.8}}$

Speeds

The interaction between the rolling element and the races in this type of bearing has a straight effect on performance. The angular velocity of the center, the rolling velocity at the point contact and centrifugal force for a rolling element respectively are:

$$\omega_c = \frac{R_{in}}{R_{in} + R_{out}}\omega$$

$$U_r = \omega_c R_{out}$$

$$F_c = m\omega_c^2(R_{in} + r)$$

Where ω is angular velocity of the bearing, r is the element radius and m is the mass while R_{in} and R_{out} are inner and outer ring radiuses.

Limiting temperatures are reached at different speed depending on the heat generated by friction loss and other factor such as applied loads, lubrication and bearing features. Hence it is crucial to estimate reference speed and operating temperature in order to predict bearing performance. The ISO 15312 standards present thermal speed ratings for rolling bearings under some specific conditions. When loads and oil viscosity values exceed the standards references the frictional heat loss increase what affects the reference speed of the bearings. Based on these standards, SKF established relations in order to adjust the reference speeds influenced by high viscosity or loads. In fact the adjusted speed for oil lubrication can be calculated as following:

$$n_{ar} = n_r f_p f_v$$

Where n_{ar} is adjusted reference speed [r/min], n_r = nominal reference speed [r/min] f_p is adjustment factor for bearing load P and P_v is adjustment factor for oil viscosity. While for grease lubrication the equation is,

$$n_{ar} = n_r f_p \frac{f_{v\ actual\ base\ oil\ viscosity}}{f_{v\ ISO\ VG150}}$$

It is also important to set limiting speed for bearings because even under favorable conditions limiting values should not be exceeded. Generally product data specifies the reference and limiting speeds and provide diagrams for speed adjustment factors.

Failure and Wear

In addition to excessive thrust load, rolling element bearings can fail due interruption or inappropriate supply of the lubricant since in some applications this can start immediate damage. Besides thermal imbalance is harmful because inadequate thermal gradient between the outer and the inner races can result in additional radial preloading. This imbalance may also be the result of excessive sliding what may cause contact if film thickness is not redefined. Outer elements can also generate higher temperature than the bearing what can affect the lubricant and the bearing element so the design should take in consideration the dissipation of this heat away from the bearing. Also if radial clearance between the two raceways is lost or the bearing is misaligned, the load may increase and fracture occurs at the cage. In applications where bearings are exposed to electrical current, tiny holes will eventually form in the bearing surfaces. Moreover, adhesive wear is common in rolling element bearings especially by start-stop operations while chemical corrosive wear could occur as result of interaction with the outer environment or the lubricant. Impact wear is probable with applications where sudden impacts are endured. Moreover fatigue wear is the most common for these bearings especially for heavy-loaded ones.

Lubrication

Generally for efficient lubrication the minimum film thickness must exceed the equivalent surface roughness at the contact. Lubrication is de-fined by ratio Λ of film thickness to equivalent roughness of the contacting surfaces. For rolling bearings this ratio is usually between 3 and 5. In fact rolling bearing lubrication operates under elasto-hydrodynamic regime. Minimum thickness for ball bearing can be calculated from the equation of point contact, while roller bearing is defined by line contact minimum thickness equation. Attention should be kept while determining the equivalent radius of curvatures between the rolling elements and races. The following figure presents a summary of sign convention for rolling bearings radii.

Usually the range of elastohydrodynamic film thickness lubrication is between 0.1 and few micrometers with bearings operating under normal conditions. Since the geometry of the rolling elements is not perfect then contact can occur, thus additives are used to enhance the lubricant performance. There are different lubrication methods used for rolling elements bearings depending on the application. The most used mode is grease lubrication because it is the less expensive and the simplest. The grease is transferred away from the rolling elements path and a portion of the fluid flow into the raceway ensuring the lubrication function. Grease is held using seals or shields. However grease does not react to operating speed change and the selection of the grease type is based on the working conditions. The grease operating life depends on leakage rate of the oil out of the grease and by degradation or evaporation. The period of relubrication is estimated using previous experience or similar model of bearings. Generally grease is widely used because of the light weight, the simple design and the facile maintenance. Solid film lubrication is used for extreme conditions like very low or very high temperatures and also low pressure. The film can be powder, bonded or transfer films. In most of bearing using this lubrication, the cage is made from PTFE. Rolling elements pick a PTFE film while contacting the cage pocket and the bearing is cooled. It is mostly used for aerospace industry. From another side jet lubrication is mostly used with extreme bearing speeds to immediately remove the heat generated. The performance of this technique depends on the supplying nozzles and the flow rate of lubricant. The centrifugal forces participate in the heat decrease by moving the lubricant through the bearing.

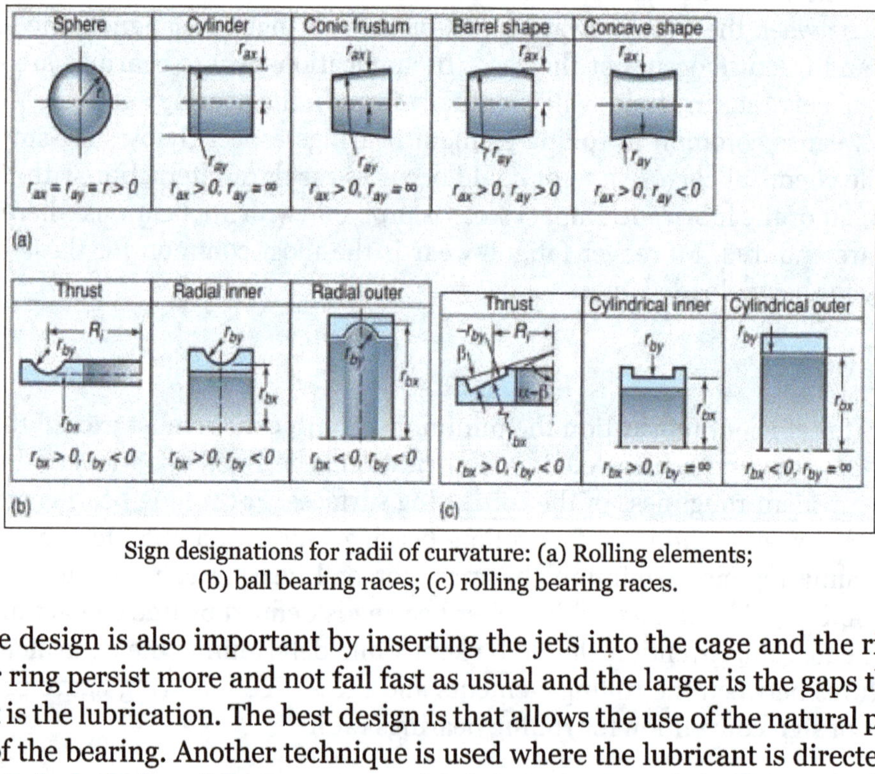

Sign designations for radii of curvature: (a) Rolling elements; (b) ball bearing races; (c) rolling bearing races.

The cage design is also important by inserting the jets into the cage and the ring gaps so outer ring persist more and not fail fast as usual and the larger is the gaps the more efficient is the lubrication. The best design is that allows the use of the natural pumping ability of the bearing. Another technique is used where the lubricant is directed under and through the holes of the inner race. This lubrication using under-race passages is more successful in operation where cooling the inner race is needed since jet lubrication failed because of lubricant outwards conduction by centrifugal forces. This technique is difficult to apply with small bores because of the limited space but it is very efficient with large bores such used in aircraft turbine engines. The difficulty of this technique has limited the use of this lubrication to special applications since it requires holes through the inner race of the bearing what may lead to fast fracture and failure. Besides, mist lubrication is common with these bearings. Actually, the technique suggests the suspension of oil droplets in air as a fog to be carried into the bearing. The condensation of these particles will result to lubrication. Nozzles are used to control the speed of the mist. However the oil mist passes only once through the bearing and then discarded. This lubrication mode is very effective with high speed and precision tools in machinery for example in helicopter turbine engines. The following figures shows the operating principles of some lubrication techniques.

Bearing Life

Lundberg-Palmgren Bearing Life Theory

Weibull distribution is a probability model that estimates the distribution of a bearing life based on statistical data. The theory established by Weibull was developed by Lundberg and Palmgren based on the influence of the rolling element-raceway contact normal stress, on the fatigue life of raceways. Based on Weibull probability distribution of metal fatigue the concluded the basic theory of the stochastic dispersion. They assumed that the logarithm of the reciprocal of the probability of survival S could be expressed as a power function of orthogonal shear stress x_0, life N, depth to the maximum orthogonal shear stress Z_0, and stressed volume V. That is,

$$\ln\frac{1}{S}=\frac{\tau_0^c N^e}{z_0^h}V$$

Including a as the semi-major axis of the Hertzian contact ellipse, and l as the length of the running track of the race where V= alz_0 then,

$$\ln\frac{1}{S}=\frac{\tau_0^c N^e al}{z_0^{h-1}}$$

By substituting the Hertzian contact parameters (in terms of the applied load and contact geometry), they obtained the load-life relationship for rolling bearings. This can be expressed in its final form, called basic rating life, in a very simple way:

$$L_{10}=(\frac{C_r}{P_{eq}})^p$$

Where C is the bearing basic dynamic load rating, p is factor that depends on the bearing geometry (3 for ball bearings and 10/3 for roller bearings), and P is the equivalent bearing load and L_{10} is the rated life at 1 million inner-race revolutions with 90% probability,

$$C_r=f_{cm}\left(icos\alpha\right)^{0.7}Z^{2/3}d^{1,8}$$

Where Cr is the basic dynamic load rating, f_{cm}, is the bearing geometry and material coefficient per ANSI/ABMA standards, i is the number of rolling-element rows, α is the bearing contact angle, Z is the number of elements per row, and d is the ball diameter, while

$$P_{eq} = XF_r + YF_a$$

Where F_r and F_a are, respectively, the radial and axial loads applied to the bearing and X and Y are factors calculated by Lundberg and Palmgren to provide the proper equivalent load based on the ratio F_a/F_r. For pure radial loads, P_{eq} equals F_r.

The Palmgren theory can be proved starting with assumption that any material has a certain fatigue limit that can be infinite under limited load level. So the curve life is asymptotic, also the material must yield to a finite load. With the further assumption that the curve has the profile of an exponential function the following function can be established:

$$k = C(an + e)^{-x} + u$$

Where k is the specific load or Stribeck's constant, C is the material constant, a is the number of load cycles during one revolution at the point with the maximum load exposure, n is the number of revolutions in millions, e is the material constant that is dependent on the value of the elasticity or fracture limit, u is the fatigue limit, and x is an exponent (always close to 1/3).

The previous equation can be written as,

$$\text{Life (millions of stress cycles)} = (\frac{C}{k-u})^3 - e\frac{n!}{r!(n-r)!}$$

if e is zero and eliminating the fatigue limit the relation becomes,

$$\text{Life}\left(\text{millions of race rev}\right) = (\frac{CZd^2/5}{Q})^3$$

Recalling the equation of basic load C_r, using f_{cm} = C/5 while P_{eq} = Q and α = 0 finally the result is,

$$L = (\frac{C_r}{P_{eq}})^3$$

ISO Life

The DIN ISO 281 standards state the use of the basic life rating for life determination. And also if additional information known about the material and the lubrication more accurate raring can be used called adjusted life rating L_{na} as follows:

$$L_{na} = a_1 a_2 a_3 L_{10}$$

Where a_1 is life adjustment factor for requisite reliability other that 90%, a_2 is adjustment factor for materials characteristics and a_3 is factor for operating conditions. However in 2007 ISO improved this to new rating called expanded adjusted rating life defines as,

$$L_{nm} = a_1 a_{iso} L_{10} \left(\text{in } 10^6 \text{ revolutions}\right)$$

Where a_{iso} is life adjustment factor for operating conditions, a1 was redefined.

The a_{iso} factor takes into consideration loads, lubrication and lubricant features, residual stress and fatigue limit of the material and operating conditions. This factor can be calculated as,

$$a_{iso} = f(\frac{e_c C_u}{P} k)$$

Where e_c is a life adjustment factor for contamination, C_u is the fatigue limit load, P is the equivalent dynamic bearing load, k is the viscosity ratio (For $k > 4$, calculation should be carried out using $k = 4$. For $k < 0{,}1$, this calculation method cannot be used).

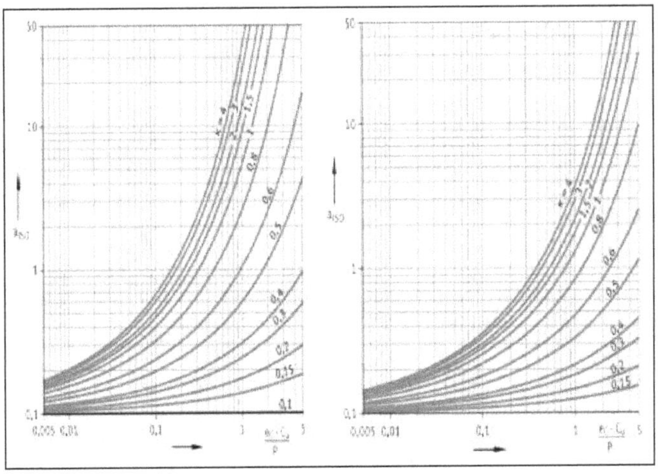

a_{iso} values for radial and axial roller bearing.

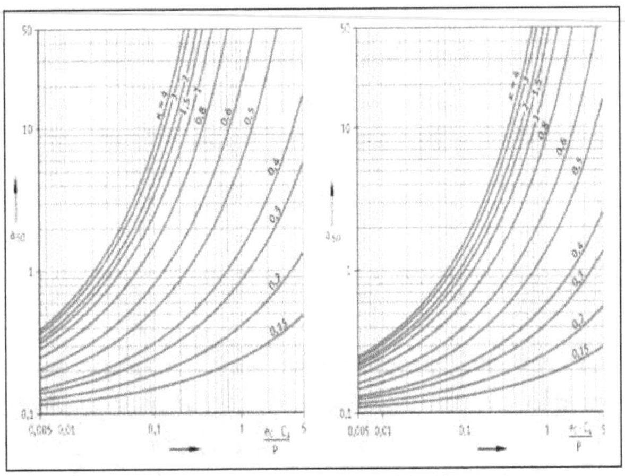

a_{iso} values for radial and axial ball bearing.

Involute Gears

Gears Contact and Stress

The study of tribological behavior of involute gears is based on the understanding of the contact generated between the mating gears. In order to establish a concluding analysis some assumptions should be respected. In fact involute teeth should be perfectly shaped and equally spaced, also having equal share of load while contacting and constant pressure distribution. Analysis of mating gears as cylinder is possible but not precise since the radii of curvature are variable because of the teeth profile. Furthermore, the deformation of teeth is hard to predict especially with the introduction of lubricants. As shown in the following figures, the mesh of 2 involute teeth generates different successive line of contact and the succession of these lines forms the line of action in the transversal direction.

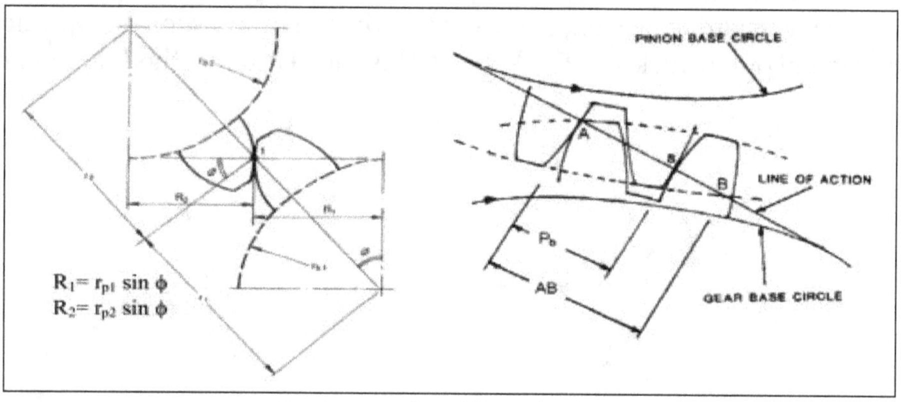

Involute teeth and line of action.

The maximum Hertzian contact for involute teeth is then,

$$\sigma_0 = \sqrt{\frac{W(1+r_{p1}/r_{p2})}{f\pi r_{p1}[(1-v_1^2)/E_1(1-v_1^2)/E_2]\sin\phi}}$$

Where W is the normal load, E_1 and E_2 are modulus, v_1 v_2 are Poisson ratio, rp_1 rp_2 are gears diameters, f is gear width and φ is the pressure angle. The contact is quantified using the contact ratio to define the number of contacting teeth for spur gear at any time as follows:

$$C_r = \frac{q_t}{p_b} = \frac{q_t p_d}{\pi \cos\varphi}$$

Where p_b is base pitch and p_d is diameter pitch and q_t is length of arc action. Generally for efficient gears this ratio should be less than 1,2. For other types of gears the additional parameters should be included to evaluate this ratio such as working pressure angle, center distance and corrected radii values, and standards should be followed depending on the gear type. Furthermore the design of gears is based on the evaluation of 2 important parameters which are contact stress number S_c and bending stress number S_b defining the maximum stress values for contacting teeth. For different types of gears, the calculation of these parameters is set by different standards such ISO 6336 and SFS 4790. The AGMA standards2001-C95 for spur and helical gears use the following equation:

$$S_c = \frac{f_t p_d}{fJ} k_o k_s k_m k_b k_v$$

$$S_b = C_p \sqrt{\frac{f_t k_o k_s k_m k_b k_v}{fID_p}}$$

Where f_t is tangential force on gear teeth, p_d diametral pitch of the gear, f face width of the gear, I and J are geometry factors for contact stress and bending stress, k_o overload factor, k_s seize factor, k_b rim thickness factor, km load distribution factor, k_v dynamic factor and C_p is elastic coefficient of the gear material and D_p pitch diameter of the pinion. Standards should be followed for different factors and for calculations depending of the gear types.

Gears Friction

Friction Components

The friction in gears exists between contacting teeth as result of relative sliding. The analysis of friction in this case should be taken into 2 stages, start point-pitch point and pitch point-end point. There is no friction at the pitch point location. As shown in the following figure, the friction force component direction changes at this point due to change in sliding direction.

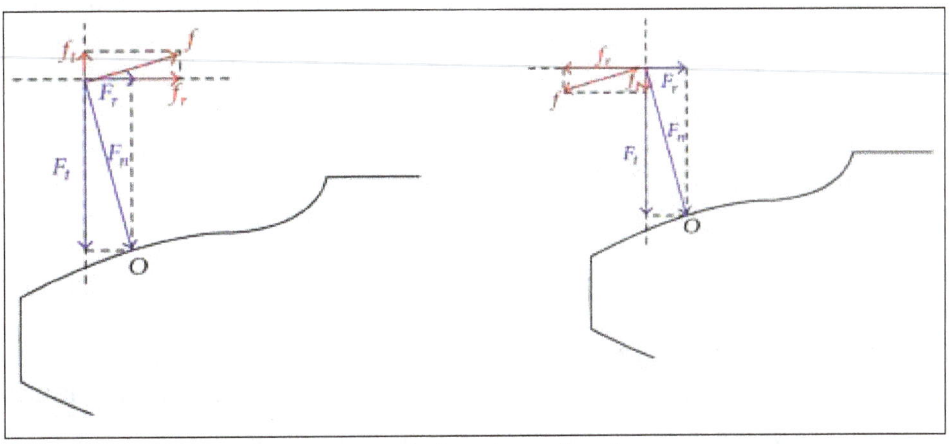

Force system during approach and recess.

For both driven and driving gears, the friction force is always defined as the product of the net normal force by coefficient of sliding friction μ. However the vertical input force F_v and the horizontal input force F_h are defined as following:

During approach:

$$F_v = F_t - f_t$$
$$F_h = F_r + f_r$$

During recess:

$$Fv = F_t + f_t$$
$$F_h = F_r - f_r$$

This phenomena shows the rolling/friction duality and has great effect on bending stress behavior, In fact bending stress is amplified during approach and weaken during recess by friction effect. Hohn et al developed the following expression for friction co-efficient in gears:

$$\mu = 0.048\left(\frac{F_t}{\rho_c b v_{\Sigma c}}\right)^{0.2} v_0^{-0.05} R_a^{0.25} X_L X_{SC}$$

Where F_t is the normal force in transverse direction, b is the face width, $v_{\Sigma c}$ is the sum velocity at pitch point, ρ_c is the equivalent curvature radius at the same point, v_0 is the kinetic viscosity, R_a is the composite surface roughness, X_L is a correction factor for additives and X_{SC} is a correction factor for coating.

Friction Loss

Power losses in gears are mainly due friction and churning and generally for different types of gears the friction losses are about 1 % form total input power. For gears it is common to define the friction losses as efficiency percentage. There are different methods to estimate these fiction losses. Actually, if the work done by pinion and gear is to be considered and if the friction coefficient is constant the following relation are found:

$$f_{/\min} = \frac{F_n \omega R_b}{F_t + F_r}\left[\frac{\mu'}{2}(\beta_t^2 + \beta_r^2)(1 + \frac{1}{i})\right]$$

$$\text{efficiency} = 1 - \left[\frac{1 + (1/i)}{\beta_t + \beta_r}\right]\frac{\mu'}{2}(\beta_t^2 + \beta_r^2)$$

Where f/min is the friction loss per minute, i s the gear ratio, ω is the angular velocity of driving gear, μ' is average friction coefficient, R_b is base radius of the driver, β_t and β_r is the arc of approach and recess of the driver. Different speeds, loads, lubricants can change the friction coefficient from constant to variable then the relations shift to,

$$f_{/\min} = \frac{F_n \omega R_b}{F_t + F_r}\left[(1 + \frac{1}{i})(\frac{\mu_a}{2}\beta_t^2 + \frac{\mu_r}{2}\beta_r^2)\right]$$

$$\text{efficiency} = 1 - [\frac{1+(1/i)}{\beta_t + \beta_r}(\frac{\mu_a}{2}\beta_t^2 + \frac{\mu_r}{2}\beta_r^2)]$$

Where μ_a and μ_r are average friction coefficient during approach and recess. If the losses are to be calculated using energy calculation approach the results are:

$$\text{Effeciency} = 100 - \text{P} = 1 - \frac{50\mu(H_s^2 + H_s^2)}{\phi(H_s + H_t)}$$

For spur gear and helical gears:

$$H_t(\frac{\text{R}_g \pm 1}{\text{R}_g})[\sqrt{((\frac{r_o}{r_p})^2 - \cos^2\alpha)} - \sin\alpha]$$

$$H_s = (\text{R}_g \pm 1)[\sqrt{((\frac{R_o}{R_p})^2 - \cos^2\alpha)} - \sin\alpha]$$

Table: Φ Parameter Calculation.

Gear Type	Φ	sign of R_g term
External spur	$\cos\alpha$	(R_g+1)
Internal Spur	α	(R_g-1)
Single Helical	$\cos\varphi n/\cos^2\beta \ \tan\varphi_n = \tan\alpha.\cos\beta \ \beta = \text{helix angle}$	(R_g+1)

For worm gear the following formula is used:

$$\text{effeciency} - \text{worm} = \frac{\cos\alpha_n - \mu\tan\gamma}{\cos\alpha_n + \mu\cot\gamma}$$

Where R_g is Gear ratio, R_o is Outside Diameter of gear (m), r_o is Outside Diameter of pinion(m), R_p is Pitch Diameter of gear (m), r_p is Pitch Diameter of pinion(m), E is efficiency of gear pair (%), P is Power loss of gear pair as a percentage of input power (%), α_n is normal pressure angle, v_s is gear surface sliding velocity (m/s), α is pressure angle, β is helix angle (deg), γ is worm lead angle (deg) and μ is coefficient of friction.

Wear and Failure Modes for Gears

ISO 10825:1995 and ANSI/AGMA 1010E95 defines failure and wear modes for different types of gears. There are many gears failures modes that can manifest as wear, deformation, fracture, cracking or fatigues. Usually these modes are classified under lubrication related or nonlubrication related. Gear teeth flanks are subject to failure caused by wear, pitting and scuffing while gear root fillets brake after fatigue. Bending fatigue is the most common type of fatigue failure caused usually by imperfections on the teeth surface or exceeding of the bending contact number. It is initiated by cracks that propagate under loading until fracture. Contact fatigue is also important; it is usually triggered when rolling or sliding loadings cause friction that changes the distribution of the contact interface stress. When contact stress number is exceeded cracks are formed at the gear teeth until propagation to the surface and possible pitting. Thermal fatigue can occur at high friction or high temperature working environment. Moreover, impact should be also considered, actually it can be tooth-bending impact by fast removing of gear tooth, tooth-shear when high impact is endured at short contacting time or tooth chipping when an external force alter the stress features of the teeth. Pitting is very common phenomena in gears and it exists even with lubrication and without contact.

In fact lubricant may enhance pitting under high pressures by forcing lubricant into cracks what lead to their spread. Smooth surfaces usually present less amount of pitting. For through-hardened gears, pitting occurs at the surface of the teeth, it starts at the pitch line and progress towards dedendum which is usually the first to manifest pitting. Usually this surface originated pitting risk is evaluated by comparing the predicted stress in teeth to the contact stress number. Surface-hardened or case carburized teeth usually develop subsurface originated pitting. Cracks are initiated it the case/core interface and after coalescence lager volume in teeth is harmed. Subsurface pitting is frequent in slow motion final drives using spur, helical bevel or hypoid gears. Deepening the case interface and decreasing loads are efficient ways to limit this kind of pitting. Generally, subsurface pitting should be considered while designing by evaluation of the half width of Hertzian contact and the shear stress below surface to secure the case-depth. Abrasive wear can occur with gears especially if hard abrasive particles are inserted into contact area. These particles can be generated by lubricant or by neighboring elements such seals or bearing. Also chemical wear can be caused by chemical elements in operating area. However adhesive wear is not common and only manifest when particles from other failed components are impinged into the gear teeth.

Scuffing is most frequent excessive damage of gears wear characterized by local weld between surfaces and it is mainly due to lubrication failure at high temperature. Scuffing exhibit torn markings in sliding direction and it is hard to be controlled. In fact, with sliding friction generates additional stress that can start plastic deformation in the surface layer and hence welding is possible. So it is the result of friction coefficient increase and it is usually localized away from the pitch line. Beside natural ductile films, scuffing can be reduced by appropriate material selection by using chemically similar material with different hardness to keep at least one intact and also it is know that material having low solid solubility are less sensitive to weld. However, the best way to limit the scuffing risk is to ensure adequate lubrication. In addition to film thickness, critical temperature at the contact zone is very important factor affecting scuffing. This explained by the failure of lubricating oil at high temperature and therefore eventual gear failure. Therefore a limit temperature called flash temperature T_f is evaluated and then compared to the maximum lubricant temperature. It is calculated as follows:

$$T_f = T_b + \left(\frac{1.25}{1.25 - R_a}\right) G_c \left(\frac{P_e}{b}\right)^{0.75} \left(\omega_1^{0.5} \frac{m^{0.25}}{1.094}\right)$$

Where T_b is gear bulk temperature, b is face width, m is module, R_a is surface finish, P_e is to be taken as full load on teeth (60% only for perfect geometry) and G_c is geometry constant.

Gears Lubrication

Lubrication Regimes of Gears

Gears are generally operating under one of three regimes depending mainly on the operating speed. In fact, at low speeds boundary lubrication is the dominant lubrication regime. Eleastohydrodynamic films do not exist and gear protection from wear and failures is done through absorption of very thin surface film built by lubricant and additives, the friction coefficient in this regime is about 0.1 to 0.2. At moderate velocities, mixed lubrication is the dominant regime when the speed allows the formation of film that does not separate totally the contacting teeth, the development of this film reduces friction and wear rates, friction values are between a range of 0,05 to 0.1. However at sufficient speeds elastohydrodynamic regime is the ruling regime after formation of an elastohydrodynamic thick film that ensure total separation of the contacting teeth and hence the friction is reduced only the share of this film having a range between 0.01 and 0.04, while the wear rate is neglected and only persists if abrasive particles in the lubricant initiate some holes into lubricated surfaces. Also, the average load intensity on surface known as factor Q contributes at the range of different gears lubrication regimes. The following figure shows the regime limits in function of factor at different velocities.

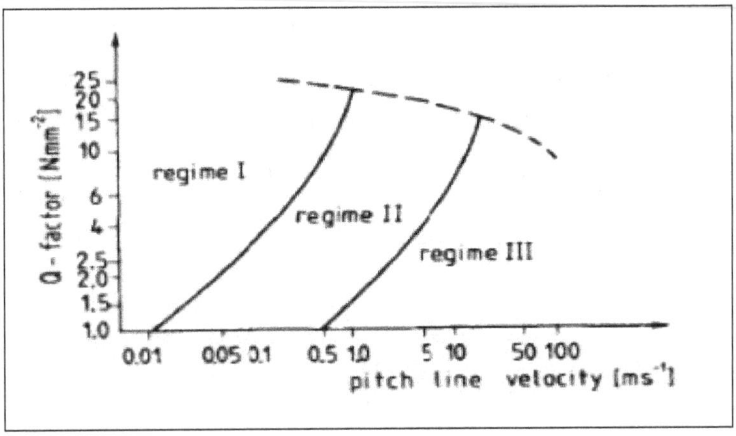

Regimes range for different loads intensity and pitch line velocities.

Minimum Film Thickness

For fast estimation, the lubrication ratio Lambda can be used for film thickness. For gears usually values greater than 1.2 are sufficient. Also, theory of line contacts and replacement of gear teeth by cylinders can be used but not very efficient. However modern design of gars use a standard formula for spur and helical gear as following:

$$h = 44.6C_f \frac{L_f V_f}{p_f}$$

$$C_f = \text{curvative factor} = \frac{C\sin\phi}{\cos^2\psi}\frac{m}{(m+1)^2}$$

$$L_f = \text{lubricant factor} = (\alpha E')^{0.54}$$

$$p_f = \text{loading factor} = (\frac{P_t}{bE'A})^{0.13}$$

$$V_f = \text{Velocity} = [\frac{\mu\eta_0}{AE'}]^{0.7}$$

$$\mu = \text{rolling velocity} = \frac{\pi n_p d_p \sin\phi}{60}$$

$$E' = \text{effective modulus} = \frac{\pi}{2}\frac{E}{(1-v)^2}$$

Where c is center distance, ψ is helix angle, ϕ is pressure angle, m is module,α is pressure –viscosity coefficient,η_0 is viscosity at operating temperature, d_p is pitch diameter, n_p is pinion velocity in rpm, P_t is tangential load on tooth , a is loaded area and b is face width.

References

- Bearing-materials, basics-design: machinedesign.com, Retrieved 08 July, 2020
- Journal-bearing-lubrication-779: machinerylubrication.com, Retrieved 28 April, 2020
- Tilting-pad-thrust-bearings-906: machinerylubrication.com, Retrieved 05 August, 2020
- Batchelor, A. and Stachowiak, G.2005. Engineering tribology. Oxford: Elsevier
- Bhushan, B and Nosonovsky, M.2008. Multiscale Dissipative Mechanisms and Hierarchical Surfaces. New-York: Springer

Tribological Aspects of Gears

5

- **Spur Gears**
- **Gear Failure and its causes**
- **Monitoring System of Gears**

There are many aspects that are dealt within the tribology of gears such as spur gears, gear configurations, contact mechanics, monitoring system of gears, gear failure and its management, etc. The topics elaborated in this chapter will help in gaining a better perspective of the tribological aspects of gears.

Spur Gears

Spur gears are a type of cylindrical gear, with shafts that are parallel and coplanar, and teeth that are straight and oriented parallel to the shafts. They're arguably the simplest and most common type of gear – easy to manufacture and suitable for a wide range of applications.

The teeth of a spur gear have an involute profile and mesh one tooth at a time. The involute form means that spur gears only produce radial forces (no axial forces), but the method of tooth meshing causes high stress on the gear teeth and high noise production. Because of this, spur gears are typically used for lower speed applications, although they can be used at almost any speed.

An involute gear tooth has a profile that is the involute of a circle, which means that as two gears mesh, they contact at a single point where the involutes meet. This point moves along the tooth surfaces as the gears rotate, and the line of force (known as the line of action) is tangent to the two base circles. Thus, the gears adhere to the fundamental law of gearing, which states that the ratio of the gears' angular velocities must remain constant throughout the mesh.

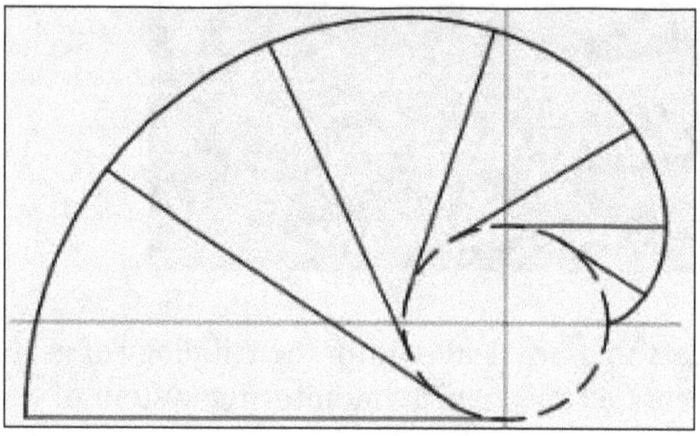

Spur gears can be made from metals such as steel or brass, or from plastics such as nylon or polycarbonate. Gears made of plastic produce less noise, but at the expense of strength and loading capability. Unlike other gear types, spur gears don't experience high losses due to slippage, so they generally have high transmission efficiency. Multiple spur gears can be used in series (referred to as a gear train) to achieve large reduction ratios.

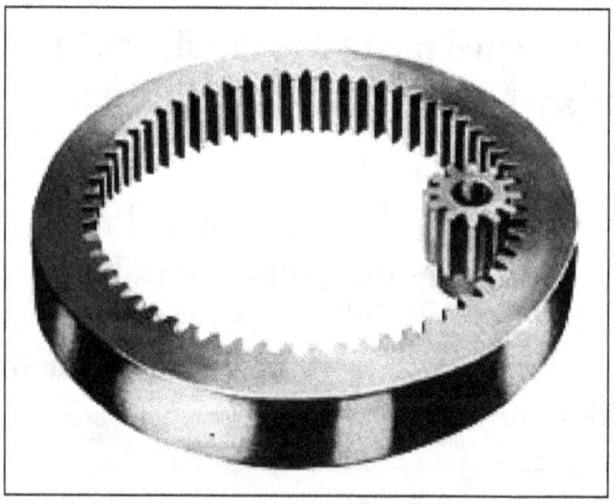

An external gear (with teeth on the outside of the cylinder) rides on the inside surface of an internal gear.

There are two primary types of spur gears: external and internal. External gears have teeth that are cut on the outside surface of the cylinder. Two external gears mesh together and rotate in opposite directions. Internal gears, in contrast, have teeth that are cut on the inside surface of the cylinder. An external gear sits inside the internal gear, and the gears rotate in the same direction. Because the shafts are positioned closer together, internal gear assemblies are more compact than external gear assemblies. Internal gears are primarily used for planetary gear drives.

Spur gears are generally seen as best for applications that require speed reduction and torque multiplication, such as ball mills and crushing equipment. Examples of high-speed applications that use spur gears – despite their high noise levels – include consumer appliances such as washing machines and blenders. And while noise limits the use of spur gears in passenger automobiles, they are often used in aircraft engines, trains, and even bicycles.

Contact Mechanics

Stresses in a contact area loaded simultaneously with a normal and a tangential force. Stresses were made visible using photoelasticity.

Contact mechanics is the study of the deformation of solids that touch each other at one or more points. A central distinction in contact mechanics is between stresses acting perpendicular to the contacting bodies' surfaces (known as the normal direction) and frictional stresses acting tangentially between the surfaces. Normal stresses are caused by applied forces and by the adhesion present on surfaces in close contact even if they are clean and dry.

Contact mechanics is part of mechanical engineering. The physical and mathematical formulation of the subject is built upon the mechanics of materials and continuum mechanics and focuses on computations involving elastic, viscoelastic, and plastic bodies in static or dynamic contact. Contact mechanics provides necessary information for the safe and energy efficient design of technical systems and for the study of tribology, contact stiffness, electrical contact resistance and indentation hardness. Principles of contacts mechanics are implemented towards applications such as locomotive wheel-rail contact, coupling devices, braking systems, tires, bearings, combustion engines, mechanical linkages, gasket seals, metalworking, metal forming, ultrasonic welding, electrical contacts, and many others. Current challenges faced in the field may include stress analysis of contact and coupling members and the influence of lubrication and material design on friction and wear. Applications of contact mechanics further extend into the micro- and nanotechnological realm.

Classical Solutions for Non-adhesive Elastic Contact

The theory of contact between elastic bodies can be used to find contact areas and indentation depths for simple geometries. Some commonly used solutions are listed below. Solutions for multitude of other technically relevant shapes, e.g. the truncated cone, the worn sphere, rough profiles, hollow cylinders, etc.

Contact between a Sphere and a Half-space

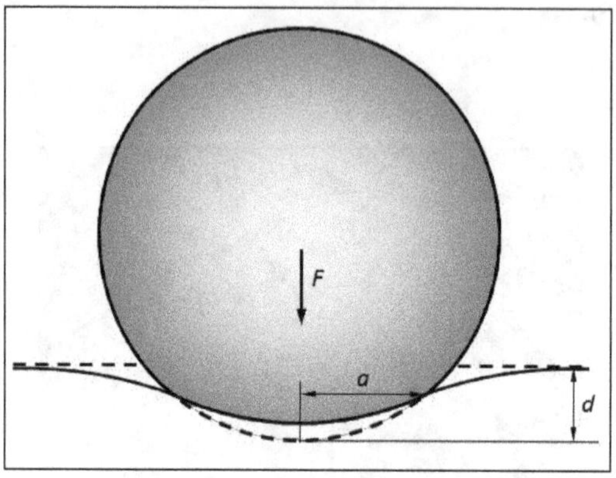

Contact of an elastic sphere with an elastic half-space.

An elastic sphere of radius R indents an elastic half-space to depth d, and thus creates a contact area of radius,

$$a = \sqrt{Rd}$$

The applied force F is related to the displacement d by,

$$F = \frac{4}{3} E^* R^{\frac{1}{2}} d^{\frac{3}{2}}$$

where,

$$\frac{1}{E^*} = \frac{1-\nu_1^2}{E_1} + \frac{1-\nu_2^2}{E_2}$$

and E_1, E_2 are the elastic moduli and ν_1, ν_2 the Poisson's ratios associated with each body.

The distribution of normal pressure in the contact area as a function of distance from the center of the circle is,

$$p(r) = p_0 \left(1 - \frac{r^2}{a^2}\right)^{\frac{1}{2}}$$

where is the maximum contact pressure given by,

$$p_0 = \frac{3F}{2\pi a^2} = \frac{1}{\pi}\left(\frac{6FE^{*2}}{R^2}\right)^{\frac{1}{3}}$$

The radius of the circle is related to the applied load F by the equation,

$$a^3 = \frac{3FR}{4E^*}$$

The depth of indentation d is related to the maximum contact pressure by,

$$d = \frac{a^2}{R} = \left(\frac{9F^2}{16E^{*2}R} \right)^{\frac{1}{3}}$$

The maximum shear stress occurs in the interior at $z \approx 0.49a$ for $\nu = 0.33$.

Contact between Two Spheres

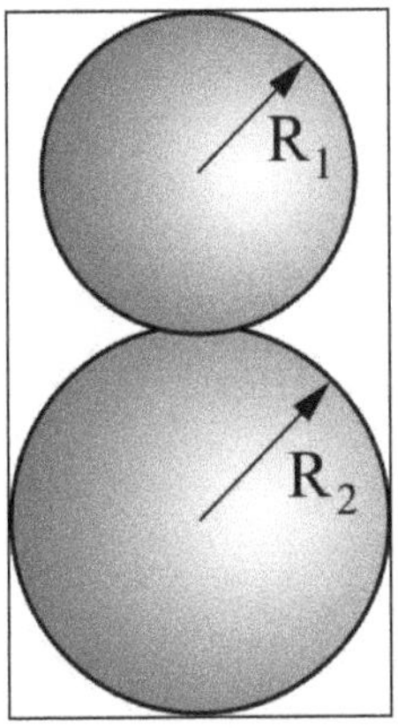

Contact between two spheres.

For contact between two spheres of radii R_1 and R_2, the area of contact is a circle of radius a. The equations are the same as for a sphere in contact with a half plane except that the effective radius R is defined as,

$$\frac{1}{R} = \frac{1}{R_1} + \frac{1}{R_2}$$

Contact between Two Crossed Cylinders of Equal Radius *R*

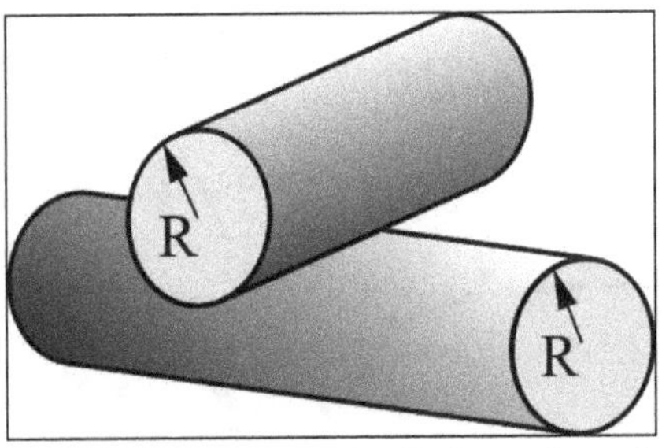

Contact between two crossed cylinders of equal radius.

This is equivalent to contact between a sphere of radius R and a plane.

Contact between a Rigid Cylinder with Flat End and an Elastic Half-space

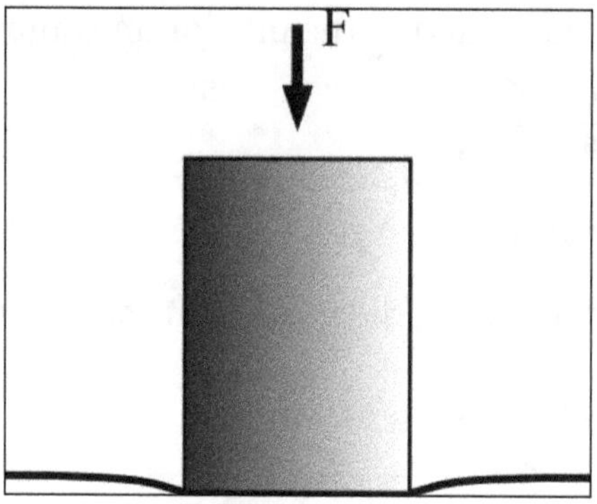

Contact between a rigid cylindrical indenter and an elastic half-space.

If a rigid cylinder is pressed into an elastic half-space, it creates a pressure distribution described by,

$$p(r) = p_0 \left(1 - \frac{r^2}{R^2}\right)^{-\frac{1}{2}}$$

where R is the radius of the cylinder and

$$p_0 = \frac{1}{\pi} E^* \frac{d}{R}$$

The relationship between the indentation depth and the normal force is given by,

$$F = 2RE^* d$$

Contact between a Rigid Conical Indenter and an Elastic Half-space

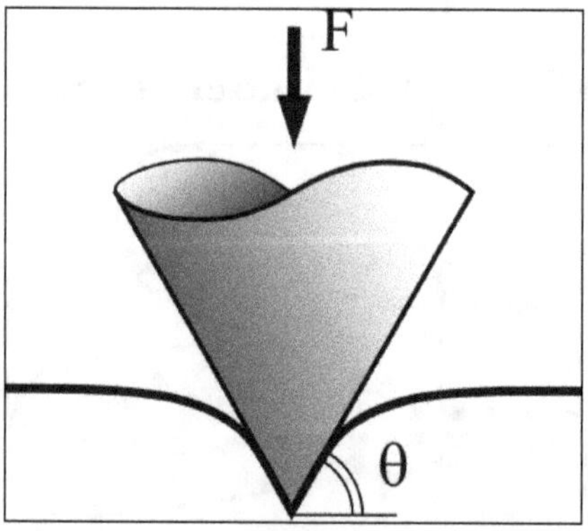

Contact between a rigid conical indenter and an elastic half-space.

In the case of indentation of an elastic half-space of Young's modulus E using a rigid conical indenter, the depth of the contact region ϵ and contact radius a are related by

$$\epsilon = a\tan(\theta)$$

with θ defined as the angle between the plane and the side surface of the cone. The total indentation depth d is given by:

$$d = \frac{\pi}{2}\epsilon$$

The total force is,

$$F = \frac{\pi E}{2\left(1-\nu^2\right)} a^2 \tan(\theta) = \frac{2E}{\pi\left(1-\nu^2\right)} \frac{d^2}{\tan(\theta)}$$

The pressure distribution is given by,

$$p(r) = \frac{Ed}{\pi a\left(1-\nu^2\right)} \ln\left(\frac{a}{r} + \sqrt{\left(\frac{a}{r}\right)^2 - 1}\right) = \frac{Ed}{\pi a\left(1-\nu^2\right)} \cosh^{-1}\left(\frac{a}{r}\right)$$

The stress has a logarithmic singularity at the tip of the cone.

Contact between Two Cylinders with Parallel Axes

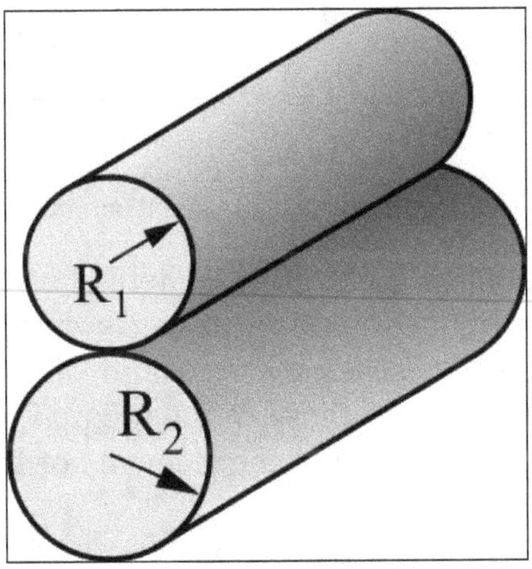

Contact between two cylinders with parallel axes.

In contact between two cylinders with parallel axes, the force is linearly proportional to the length of cylinders L and to the indentation depth d:

$$F \approx \frac{\pi}{4} E^* L d$$

The radii of curvature are entirely absent from this relationship. The contact radius is described through the usual relationship,

$$a = \sqrt{Rd}$$

with

$$\frac{1}{R} = \frac{1}{R_1} + \frac{1}{R_2}$$

as in contact between two spheres. The maximum pressure is equal to,

$$p_0 = \left(\frac{E^* F}{\pi L R} \right)^{\frac{1}{2}}$$

Bearing Contact

The contact in the case of bearings is often a contact between a convex surface (male cylinder or sphere) and a concave surface (female cylinder or sphere: bore or hemi-spherical cup).

The Method of Dimensionality Reduction

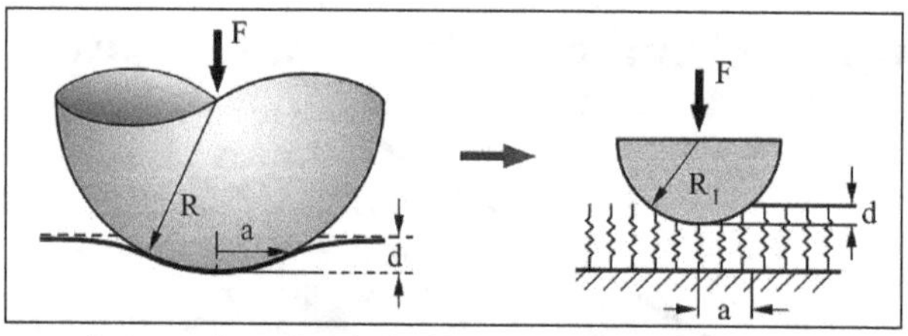

Contact between a sphere and an elastic half-space and one-dimensional replaced model.

Some contact problems can be solved with the Method of Dimensionality Reduction (MDR). In this method, the initial three-dimensional system is replaced with a contact of a body with a linear elastic or viscoelastic foundation. The properties of one-dimensional systems coincide exactly with those of the original three-dimensional system, if the form of the bodies is modified and the elements of the foundation are defined according to the rules of the MDR.

However, for exact analytical results, it is required that the contact problem is axisymmetric and the contacts are compact.

Hertzian Theory of Non-adhesive Elastic Contact

The classical theory of contact focused primarily on non-adhesive contact where no tension force is allowed to occur within the contact area, i.e., contacting bodies can be separated without adhesion forces. Several analytical and numerical approaches have been used to solve contact problems that satisfy the no-adhesion condition. Complex forces and moments are transmitted between the bodies where they touch, so problems in contact mechanics can become quite sophisticated. In addition, the contact stresses are usually a nonlinear function of the deformation. To simplify the solution procedure, a frame of reference is usually defined in which the objects (possibly in motion relative to one another) are static. They interact through surface tractions (or pressures/stresses) at their interface.

As an example, consider two objects which meet at some surface S in the (x, y)-plane with the z-axis assumed normal to the surface. One of the bodies will experience a normally-directed pressure distribution $p_z = p(x,y) = q_z(x,y)$ and in-plane surface traction distributions $q_x = q_x(x,y)$ and $q_y = q_y(x,y)$ over the region S. In terms of a Newtonian force balance, the forces:

$$P_z = \int_S p(x,y)dA; \; Q_x = \int_S q_x(x,y)dA \; ; \; Q_y = \int_S q_y(x,y)dA$$

must be equal and opposite to the forces established in the other body. The moments corresponding to these forces:

$$M_x = \int_S y q_z(x,y)dA;$$
$$M_y = \int_S -x q_z(x,y)dA;$$
$$M_z = \int_S \left[x - q_y(x,y) - Y q_x(x,y)\right]dA$$

are also required to cancel between bodies so that they are kinematically immobile.

Assumptions in Hertzian Theory

The following assumptions are made in determining the solutions of Hertzian contact problems:

- The strains are small and within the elastic limit.
- The surfaces are continuous and non-conforming (implying that the area of contact is much smaller than the characteristic dimensions of the contacting bodies).
- Each body can be considered an elastic half-space.
- The surfaces are frictionless.

Additional complications arise when some or all these assumptions are violated and such contact problems are usually called non-Hertzian.

Analytical Solution Techniques

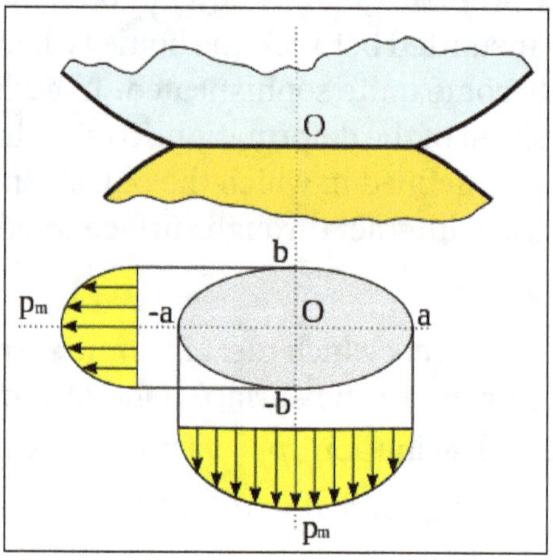

Contact between two spheres.

Analytical solution methods for non-adhesive contact problem can be classified into two types based on the geometry of the area of contact. A conforming contact is one in which the two bodies touch at multiple points before any deformation takes place (i.e., they just "fit together"). A non-conforming contact is one in which the shapes of the bodies are dissimilar enough that, under zero load, they only touch at a point (or possibly along a line). In the non-conforming case, the contact area is small compared to the sizes of the objects and the stresses are highly concentrated in this area. Such a contact is called concentrated, otherwise it is called diversified.

A common approach in linear elasticity is to superpose a number of solutions each of which corresponds to a point load acting over the area of contact. For example, in the case of loading of a half-plane, the Flamant solution is often used as a starting point and then generalized to various shapes of the area of contact. The force and moment balances between the two bodies in contact act as additional constraints to the solution.

Point Contact on a (2D) Half-plane

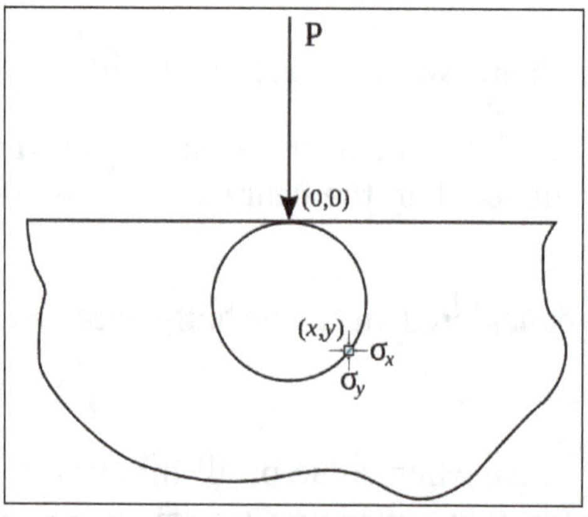

Schematic of the loading on a plane by force P at a point (0, 0).

A starting point for solving contact problems is to understand the effect of a "point-load" applied to an isotropic, homogeneous, and linear elastic half-plane, shown in the figure to the right. The problem may be either plane stress or plane strain. This is a boundary value problem of linear elasticity subject to the traction boundary conditions:

$$\sigma_{xz}(x,0)=0\ ;\ \sigma_z(x,z)=-P\delta(x,z)$$

where $\delta(x,z)$ is the Dirac delta function. The boundary conditions state that there are no shear stresses on the surface and a singular normal force P is applied at (0, 0). Applying these conditions to the governing equations of elasticity produces the result,

$$\sigma_{xx}=-\frac{2P}{\pi}\frac{x^2z}{\left(x^2+z^2\right)^2}$$

$$\sigma_{zz}=-\frac{2P}{\pi}\frac{z^3}{\left(x^2+z^2\right)^2}$$

$$\sigma_{xz}=-\frac{2P}{\pi}\frac{xz^2}{\left(x^2+z^2\right)^2}$$

for some point, (x,y), in the half-plane. The circle shown in the figure indicates a surface on which the maximum shear stress is constant. From this stress field, the strain components and thus the displacements of all material points may be determined.

Line Contact on a (2D) Half-plane

Normal Loading over a Region (*a*, *b*)

Suppose, rather than a point load P, a distributed load $p(x)$ is applied to the surface instead, over the range $a<x<b$. The principle of linear superposition can be applied to determine the resulting stress field as the solution to the integral equations:

$$\sigma_{xx}=-\frac{2z}{\pi}\int_a^b\frac{p(x')(x-x')^2\,dx'}{[(x-x')^2+z^2]^2}\ ;\ \sigma_{zz}=-\frac{2z^3}{\pi}\int_a^b\frac{p(x')dx'}{[(x-x')^2+z^2]^2}$$

$$\sigma_{xz}=-\frac{2z^2}{\pi}\int_a^b\frac{p(x')(x-x')dx'}{[(x-x')^2+z^2]^2}$$

Shear Loading over a Region (*a*, *b*)

The same principle applies for loading on the surface in the plane of the surface. These kinds of tractions would tend to arise as a result of friction. The solution is similar the above (for both singular loads Q and distributed loads $q(x)$) but altered slightly:

$$\sigma_{xx}=-\frac{2}{\pi}\int_a^b\frac{q(x')(x-x')^3\,dx'}{\left[(x-x')^2+z^2\right]^2}\ ;\ \sigma_{zz}=-\frac{2z^2}{\pi}\int_a^b\frac{q(x')(x-x')dx'}{[(x-x')^2+z^2]^2}$$

$$\sigma_{xz} = -\frac{2z}{\pi}\int_a^b \frac{q(x')(x-x')^2 dx'}{[(x-x')^2+z^2]^2}$$

These results may themselves be superposed onto those given above for normal loading to deal with more complex loads.

Point Contact on a (3D) Half-space

Analogously to the Flamant solution for the 2D half-plane, fundamental solutions are known for the linearly elastic 3D half-space as well. These were found by Boussinesq for a concentrated normal load and by Cerruti for a tangential load.

Numerical Solution Techniques

Distinctions between conforming and non-conforming contact do not have to be made when numerical solution schemes are employed to solve contact problems. These methods do not rely on further assumptions within the solution process since they base solely on the general formulation of the underlying equations. Besides the standard equations describing the deformation and motion of bodies two additional inequalities can be formulated. The first simply restricts the motion and deformation of the bodies by the assumption that no penetration can occur. Hence the gap h between two bodies can only be positive or zero,

$$h \geq 0$$

where $h = 0$ denotes contact. The second assumption in contact mechanics is related to the fact, that no tension force is allowed to occur within the contact area (contacting bodies can be lifted up without adhesion forces). This leads to an inequality which the stresses have to obey at the contact interface. It is formulated for the normal stress.

$$\sigma_n = \mathbf{t}\cdot\mathbf{n}.$$

At locations where there is contact between the surfaces the gap is zero, i.e. $h = 0$, and there the normal stress is different than zero, indeed, $\sigma_n < 0$. At locations where the surfaces are not in contact the normal stress is identical to zero; $\sigma_n = 0$, while the gap is positive; i.e., $h > 0$. This type of complementarity formulation can be expressed in the so-called Kuhn–Tucker form, viz.

$$h \geq 0, \quad \sigma_n \leq 0, \quad \sigma_n h = 0.$$

These conditions are valid in a general way. The mathematical formulation of the gap depends upon the kinematics of the underlying theory of the solid (e.g., linear or non-linear solid in two- or three dimensions, beam or shell model). By restating the normal stress σ_n in terms of the contact pressure, p; i.e., $p = -\sigma_n$ the Kuhn-Tucker problem can be restated as in standard complementarity form i.e.

$$h \geq 0, \quad p \geq 0, \quad ph = 0.$$

In the linear elastic case the gap can be formulated as,

$$h = h_0 + g + u,$$

where h_0 is the rigid body separation, g is the geometry/topography of the contact (cylinder and roughness) and u is the elastic deformation/deflection. If the contacting bodies are approximated as linear elastic half spaces, the Boussinesq-Cerruti integral equation solution can be applied to express the deformation (u) as a function of the contact pressure (p); i.e.

$$u = \int_{\infty}^{\infty} K(x-s)p(s)ds,$$

where,

$$K(x-s) = \frac{2}{\pi E^*} \ln|x-s|$$

for line loading of an elastic half space and,

$$K(x-s) = \frac{1}{\pi E^*} \frac{1}{\sqrt{(x_1 - s_1)^2 + (x_2 - s_2)^2}}$$

for point loading of an elastic half-space.

After discretization the linear elastic contact mechanics problem can be stated in standard Linear Complementarity Problem (LCP) form.

$$\mathrm{h} = \mathrm{h}_0 + \mathrm{g} + \mathrm{Cp},$$
$$\mathrm{h} \cdot \mathrm{p} = 0, \mathrm{p} \geq 0, \mathrm{h} \geq 0,$$

where C is a matrix, whose elements are so called influence coefficients relating the contact pressure and the deformation. The strict LCP formulation of the CM problem presented above, allows for direct application of well-established numerical solution techniques such as Lemke's pivoting algorithm. The Lemke algorithm has the advantage that it finds the numerically exact solution within a finite number of iterations.

Contact between Rough Surfaces

When two bodies with rough surfaces are pressed against each other, the true contact area formed between the two bodies, A, is much smaller than the apparent or nominal contact area A_0. The mechanics of contacting rough surfaces are discussed in terms of normal contact mechanics and static frictional interactions . Natural and engineering surfaces typically exhibit roughness features, known as asperities, across a broad range of length scales down to the molecular level, with surface structures exhibiting self affinity, also known as surface fractality. It is recognized that the self affine structure of surfaces is the origin of the linear scaling of true contact area with applied pressure. Assuming a model of shearing welded contacts in tribological interactions, this ubiquitously observed linearity between contact area and pressure can also be considered the origin of the linearity of the relationship between static friction and applied normal force.

In contact between a "random rough" surface and an elastic half-space, the true contact area is related to the normal force F by,

$$A = \frac{\kappa}{E^{*}h'}F$$

with h' equal to the root mean square (also known as the quadratic mean) of the surface slope and $\kappa \approx 2$. The median pressure in the true contact surface,

$$p_{\mathrm{av}} = \frac{F}{A} \approx \frac{1}{2}E^{*}h'$$

can be reasonably estimated as half of the effective elastic modulus E^{*} multiplied with the root mean square of the surface slope h'.

GW Model

Greenwood and Williamson in 1966 (GW) proposed a theory of elastic contact mechanics of rough surfaces which is today the foundation of many theories in tribology (friction, adhesion, thermal and electrical conductance, wear, etc). They considered the contact between a smooth rigid plane and a nominally flat deformable rough surface covered with round tip asperities of the same radius R. Their theory assumes that the deformation of each asperity is independent of that of its neighbours and is described by the Hertz model. The heights of asperities have a random distribution. The probability that asperity height is between z and $z+dz$ is $\phi(z)dz$. The authors calculated the number of contact spots n, the total contact area A_r and the total load P in general case. They gave those formulas in two forms: In the basic and using standardized variables. If one assumes that N asperities covers a rough surface, then the expected number of contacts is,

$$n = N\int_{d}^{\infty}\phi(z)dz$$

The expected total area of contact can be calculated from the formula,

$$A_a = N\pi R\int_{d}^{\infty}(z-d)\phi(z)dz$$

and the expected total force is given by,

$$P = \frac{4}{3}NE_r\sqrt{R}\int_{d}^{\infty}(z-d)^{\frac{3}{2}}\phi(z)dz$$

where:

- R, radius of curvature of the microasperity.
- z, height of the microasperity measured from the profile line.
- d, close the surface.
- $E_r = \left(\frac{1-\nu_1^2}{E_1} + \frac{1-\nu_2^2}{E_2} \right)^{-1}$, composite Young's modulus of elasticity.
- E_i, modulus of elasticity of the surface.
- ν_i, Poisson's surface coefficients.

They introduced standardized separation $h = d / \sigma$ and standardized height distribution $\phi^*(s)$ whose standard deviation is equal to one. Below are presented the formulas in the standardized form.

$$F_n(h) = \int_h^{\infty} (s-h)^n \phi^*(s) ds$$

$$n = \eta A_n F_0(h)$$

$$A_a = \pi \eta A R \sigma F_1(h)$$

$$P = \frac{4}{3} \eta A E_r \sqrt{R} \sigma^{\frac{3}{2}} F_{\frac{3}{2}}(h)$$

where:

- d is the separation.
- A is the nominal contact area.
- η is the surface density of asperities.
- E^* is the effective Young modulus.

Recently the exact approximants to A_r and P were published by Jedynak. They are given by the following rational formulas, which are very exact approximants to integrals $F_n(h)$. They are calculated for the Gaussian distribution of asperities,

$$F_n(h) = \frac{a_0 + a_1 h + a_2 h^2 + a_3 h^3}{1 + b_1 h + b_2 h^2 + b_3 h^3 + b_4 h^4 + b_5 h^5 + b_6 h^6} \exp\left(-\frac{h^2}{2} \right)$$

For $F(h)$ the coefficients are,

$$[a_0, a_1, a_2, a_3]$$
$$= [0.398942280401, 0.159773702775, 0.0389687688311, 0.00364356495452]$$

$$[b_1,b_2,b_3,b_4,b_5,b_6] = \begin{bmatrix} 1.653807476138, 1.170419428529, 0.448892964428, 0.0951971709160, \\ 0.00931642803836, -6.383774657279\times10^{-6} \end{bmatrix}$$

The maximum relative error is $9.93\times10^{-8}\%$.

For $F_{\frac{3}{2}}(h)$ the coefficients are,

$$[a_0,a_1,a_2,a_3] = [0.430019993662, 0.101979509447, 0.0229040629580, 0.000688602924]$$

$$[b_1,b_2,b_3,b_4,b_5,b_6] = [1.671117125984, 1.199586555505, 0.46936532151, 0.102632881122, 0.010686348714, 0.0000517200271]$$

The maximum relative error is $1.91\times10^{-7}\%$. The paper also contains the exact expressions for $F_n(h)$,

$$F_1(h) = \frac{1}{\sqrt{2\pi}}\exp\left(-\frac{1}{2}h^2\right) - \frac{1}{2}h\,\mathrm{erfc}\left(\frac{h}{\sqrt{2}}\right)$$

$$F_{\frac{3}{2}}(h) = \frac{1}{4\sqrt{\pi}}\exp\left(-\frac{h^2}{4}\right)\sqrt{h}\left((h^2+1)K_{\frac{1}{4}}\left(\frac{h^2}{4}\right) - h^2K_{\frac{3}{4}}\left(\frac{h^2}{4}\right)\right)$$

where erfc(z) means the complementary error function and $K_v(z)$ is the modified Bessel function of the second kind.

For the situation where the asperities on the two surfaces have a Gaussian height distribution and the peaks can be assumed to be spherical, the average contact pressure is sufficient to cause yield when $p_{av} = 1.1\sigma_y \approx 0.39\sigma_0$ where σ_y is the uniaxial yield stress and σ_0 is the indentation hardness. Greenwood and Williamson defined a dimensionless parameter Ψ called the plasticity index that could be used to determine whether contact would be elastic or plastic.

The Greenwood-Williamson model requires knowledge of two statistically dependent quantities; the standard deviation of the surface roughness and the curvature of the asperity peaks. An alternative definition of the plasticity index has been given by Mikic. Yield occurs when the pressure is greater than the uniaxial yield stress. Since the yield stress is proportional to the indentation hardness σ_0, Mikic defined the plasticity index for elastic-plastic contact to be,

$$\Psi = \frac{E^*h'}{\sigma_0} > \frac{2}{3}.$$

In this definition Ψ represents the micro-roughness in a state of complete plasticity and only one statistical quantity, the rms slope, is needed which can be calculated from surface measurements. For $\Psi < \frac{2}{3}$, the surface behaves elastically during contact.

In both the Greenwood-Williamson and Mikic models the load is assumed to be proportional to the deformed area. Hence, whether the system behaves plastically or elastically is independent of the applied normal force.

GT Model

The model proposed by Greenwood and Tripp (GT), extended the GW model to contact between two rough surfaces. The GT model is widely used in the field of elastohydrodynamic analysis.

The most frequently cited equations given by the GT model are for the asperity contact area,

$$A_a = \pi^2 (\eta\beta\sigma)^2 A F_2(\lambda),$$

and load carried by asperities,

$$P = \frac{8\sqrt{2}}{15}\pi(\eta\beta\sigma)^2\sqrt{\frac{\sigma}{\beta}}E' A F_{\frac{5}{2}}(\lambda),$$

where:

$\eta\beta\sigma$, roughness parameter,

A, nominal contact area,

λ, Stribeck oil film parameter, first defined by Stribeck as $\lambda = h/\sigma$,

E', effective elastic modulus,

$F_2, F_{\frac{5}{2}}(\lambda)$, statistical functions introduced to match the assumed Gaussian distribution of asperities.

The exact solutions for A_a and P are firstly presented by Jedynak. They are expressed by F_n as follows:

$$F_2 = \frac{1}{2}\left(h^2+1\right)\mathrm{erfc}\left(\frac{h}{\sqrt{2}}\right) - \frac{h}{\sqrt{2\pi}}\exp\left(-\frac{h^2}{2}\right)$$

$$F_{\frac{5}{2}} = \frac{1}{8\sqrt{\pi}}\exp\left(-\frac{h^2}{4}\right)h^{\frac{3}{2}}\left(\left(2h^2+3\right)K_{\frac{3}{4}}\left(\frac{h^2}{4}\right) - \left(2h^2+5\right)K_{\frac{1}{4}}\left(\frac{h^2}{4}\right)\right)$$

where erfc(z) means the complementary error function and $K_v(z)$ is the modified Bessel function of the second kind.

In paper one can find comprehensive review of existing approximants to $F_{\frac{5}{2}}$. New proposals give the most accurate approximants to $F_{\frac{5}{2}}$ and F_2, which are reported in the literature. They are given by the following rational formulas, which are very exact approximants to integrals $F_n(h)$. They are calculated for the Gaussian distribution of asperities,

$$F_n(h) = \frac{a_0 + a_1 h + a_2 h^2 + a_3 h^3}{1 + b_1 h + b_2 h^2 + b_3 h^3 + b_4 h^4 + b_5 h^5 + b_6 h^6} \exp\left(-\frac{h^2}{2}\right)$$

For $F_2(h)$ the coefficients are,

$$[a_0, a_1, a_2, a_3] = [0.5, 0.182536384941, 0.039812283118, 0.003684879001]$$
$$[b_1, b_2, b_3, b_4, b_5, b_6] = [1.960841785003, 1.708677456715, 0.856592986083, 0.264996791567, 0.049257843893, 0.004640740133]$$

The maximum relative error is $1.68 \times 10^{-7}\%$.

For $F_{\frac{5}{2}}(h)$ the coefficients are,

$$[a_0, a_1, a_2, a_3] = [0.616634218997, 0.108855827811, 0.023453835635, 0.000449332509]$$
$$[b_1, b_2, b_3, b_4, b_5, b_6] = [1.919948267476, 1.635304362591, 0.799392556572, 0.240278859212, 0.043178653945, 0.003863334276]$$

The maximum relative error is $4.98 \times 10^{-8}\%$.

Adhesive Contact between Elastic Bodies

When two solid surfaces are brought into close proximity, they experience attractive van der Waals forces. Bradley's van der Waals model provides a means of calculating the tensile force between two rigid spheres with perfectly smooth surfaces. The Hertzian model of contact does not consider adhesion possible. However, in the late 1960s, several contradictions were observed when the Hertz theory was compared with experiments involving contact between rubber and glass spheres.

It was observed that, though Hertz theory applied at large loads, at low loads:

- The area of contact was larger than that predicted by hertz theory.
- The area of contact had a non-zero value even when the load was removed.
- There was strong adhesion if the contacting surfaces were clean and dry.

This indicated that adhesive forces were at work. The Johnson-Kendall-Roberts (JKR) model and the Derjaguin-Muller-Toporov (DMT) models were the first to incorporate adhesion into Hertzian contact.

Bradley Model of Rigid Contact

It is commonly assumed that the surface force between two atomic planes at a distance from each other can be derived from the Lennard-Jones potential. With this assumption,

$$F(z)=\frac{16\gamma}{3z_0}\left[\left(\frac{z}{z_0}\right)^{-9}-\left(\frac{z}{z_0}\right)^{-3}\right]$$

where F is the force (positive in compression), 2γ is the total surface energy of both surfaces per unit area, and z_0 is the equilibrium separation of the two atomic planes.

The Bradley model applied the Lennard-Jones potential to find the force of adhesion between two rigid spheres. The total force between the spheres is found to be,

$$F_a(z)=\frac{16\gamma\pi R}{3}\left[\frac{1}{4}\left(\frac{z}{z_0}\right)^{-8}-\left(\frac{z}{z_0}\right)^{-2}\right];\ \frac{1}{R}=\frac{1}{R_1}+\frac{1}{R_2}$$

where R_1, R_2 are the radii of the two spheres.

The two spheres separate completely when the pull-off force is achieved at $z=z_0$ at which point

$$F_a=F_c=-4\gamma\pi R.$$

Johnson-Kendall-Roberts (JKR) Model of Elastic Contact

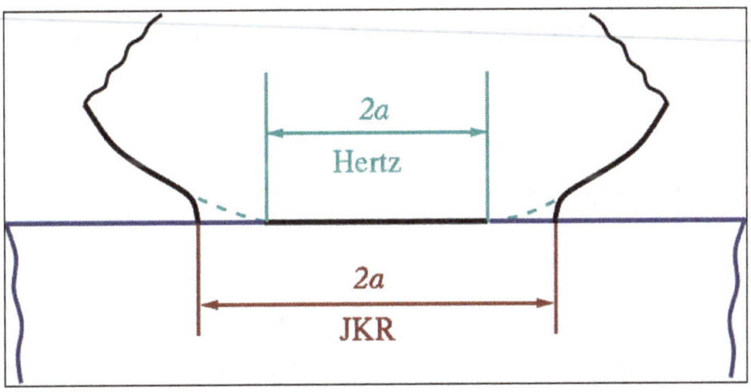

Schematic of contact area for the JKR model.

To incorporate the effect of adhesion in Hertzian contact, Johnson, Kendall, and Roberts formulated the JKR theory of adhesive contact using a balance between the stored elastic energy and the loss in surface energy. The JKR model considers the effect of contact pressure and adhesion only inside the area of contact. The general solution for the pressure distribution in the contact area in the JKR model is,

$$p(r)=p_o\left(1-\frac{r^2}{a^2}\right)^{\frac{1}{2}}+p_{o'}\left(1-\frac{r^2}{a^2}\right)^{-\frac{1}{2}}$$

Note that in the original Hertz theory, the term containing p_o' was neglected on the ground that tension could not be sustained in the contact zone. For contact between two spheres,

$$p_o=\frac{2aE^*}{\pi R};\quad p_o'=-\left(\frac{4\gamma E^*}{\pi a}\right)^{\frac{1}{2}}$$

where a is the radius of the area of contact, F is the applied force, 2γ is the total surface energy of both surfaces per unit contact area, R_i, E_i, ν_i, $i=1,2$ are the radii, Young's moduli, and Poisson's ratios of the two spheres,

$$\frac{1}{R}=\frac{1}{R_1}+\frac{1}{R_2};\quad \frac{1}{E^*}=\frac{1-\nu_1^2}{E_1}+\frac{1-\nu_2^2}{E_2}$$

The approach distance between the two spheres is given by,

$$d=\frac{\pi a}{2E^*}\left(p_o+2p_o'\right)=\frac{a^2}{R}$$

The Hertz equation for the area of contact between two spheres, modified to take into account the surface energy, has the form,

$$a^3=\frac{3R}{4E^*}\left(F+6\gamma\pi R+\sqrt{12\gamma\pi RF+(6\gamma\pi R)^2}\right)$$

When the surface energy is zero, $\gamma=0$, the Hertz equation for contact between two spheres is recovered. When the applied load is zero, the contact radius is,

$$a^3=\frac{9R^2\gamma\pi}{E^*}$$

The tensile load at which the spheres are separated (i.e., $a=0$) is predicted to be,

$$F_c=-3\gamma\pi R$$

This force is also called the pull-off force. Note that this force is independent of the moduli of the two spheres. However, there is another possible solution for the value of a at this load. This is the critical contact area a_c, given by,

$$a_c^3=\frac{9R^2\gamma\pi}{4E^*}$$

If we define the work of adhesion as,

$$\Delta\gamma = \gamma_1 + \gamma_2 - \gamma_{12}$$

where γ_1, γ_2 are the adhesive energies of the two surfaces and γ_{12} is an interaction term, we can write the JKR contact radius as,

$$a^3 = \frac{3R}{4E^*}\left(F + 3\Delta\gamma\pi R + \sqrt{6\Delta\gamma\pi RF + (3\Delta\gamma\pi R)^2}\right)$$

The tensile load at separation is,

$$F = -\frac{3}{2}\Delta\gamma\pi R$$

and the critical contact radius is given by,

$$a_c^3 = \frac{9R^2\Delta\gamma\pi}{8E^*}$$

The critical depth of penetration is,

$$d_c = \frac{a_c^2}{R} = \left(R^{\frac{1}{2}}\frac{9\Delta\gamma\pi}{4E^*}\right)^{\frac{2}{3}}$$

Derjaguin-Muller-Toporov (DMT) Model of Elastic Contact

The Derjaguin-Muller-Toporov (DMT) model is an alternative model for adhesive contact which assumes that the contact profile remains the same as in Hertzian contact but with additional attractive interactions outside the area of contact.

The radius of contact between two spheres from DMT theory is,

$$a^3 = \frac{3R}{4E^*}(F + 4\gamma\pi R)$$

and the pull-off force is,

$$F_c = -4\gamma\pi R$$

When the pull-off force is achieved the contact area becomes zero and there is no singularity in the contact stresses at the edge of the contact area.

In terms of the work of adhesion $\Delta\gamma$:

$$a^3 = \frac{3R}{4E^*}(F + 2\Delta\gamma\pi R)$$

and

$$F_c = -2\Delta\gamma\pi R$$

Tabor Parameter

In 1977, Tabor showed that the apparent contradiction between the JKR and DMT theories could be resolved by noting that the two theories were the extreme limits of a single theory parametrized by the Tabor parameter (μ) defined as,

$$\mu := \frac{d_c}{z_0} \approx \left[\frac{R(\Delta\gamma)^2}{E^{*2} z_0^3}\right]^{\frac{1}{3}}$$

where z_0 is the equilibrium separation between the two surfaces in contact. The JKR theory applies to large, compliant spheres for which μ is large. The DMT theory applies for small, stiff spheres with small values of μ.

Subsequently, Derjaguin and his collaborators by applying Bradley's surface force law to an elastic half space, confirmed that as the Tabor parameter increases, the pull-off force falls from the Bradley value $2\pi R\Delta\gamma$ to the JKR value $(3/2)\pi R\Delta\gamma$. More detailed calculations were later done by Greenwood revealing the S-shaped load/approach curve which explains the jumping-on effect. A more efficient method of doing the calculations and additional results were given by Feng.

Maugis-Dugdale Model of Elastic Contact

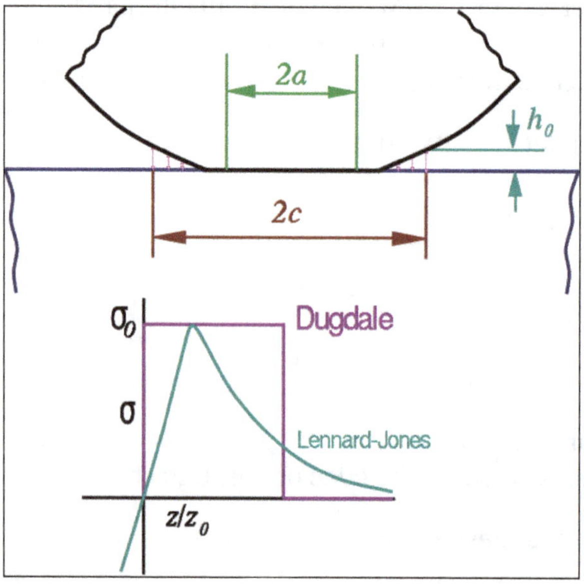

Schematic of contact area for the Maugis-Dugdale model.

Further improvement to the Tabor idea was provided by Maugis who represented the surface force in terms of a Dugdale cohesive zone approximation such that the work of adhesion is given by,

$$\Delta\gamma = \sigma_0 h_0$$

where σ_0 is the maximum force predicted by the Lennard-Jones potential and h_0 is the maximum separation obtained by matching the areas under the Dugdale and Lennard-Jones curves. This means that the attractive force is constant for $z_0 \le z \le z_0 + h_0$. There is not further penetration in compression. Perfect contact occurs in an area of radius a and adhesive forces of magnitude σ_0 extend to an area of radius $c > a$. In the region $a < r < c$, the two surfaces are separated by a distance $h(r)$ with $h(a) = 0$ and $h(c) = h_0$. The ratio m is defined as,

$$m := \frac{c}{a}.$$

In the Maugis-Dugdale theory, the surface traction distribution is divided into two parts - one due to the Hertz contact pressure and the other from the Dugdale adhesive stress. Hertz contact is assumed in the region $-a < r < a$. The contribution to the surface traction from the Hertz pressure is given by,

$$p^H(r) = \left(\frac{3F^H}{2\pi a^2}\right)\left(1 - \frac{r^2}{a^2}\right)^{\frac{1}{2}}$$

where the Hertz contact force F^H is given by,

$$F^H = \frac{4E^* a^3}{3R}$$

The penetration due to elastic compression is,

$$d^H = \frac{a^2}{R}$$

The vertical displacement at $r = c$ is,

$$u^H(c) = \frac{1}{\pi R}\left[a^2\left(2 - m^2\right)\sin^{-1}\left(\frac{1}{m}\right) + a^2\sqrt{m^2 - 1}\right]$$

and the separation between the two surfaces at $r = c$ is,

$$h^H(c) = \frac{c^2}{2R} - d^H + u^H(c)$$

The surface traction distribution due to the adhesive Dugdale stress is,

$$p^D(r)=\begin{cases} -\dfrac{\sigma_0}{\pi}\cos^{-1}\left[\dfrac{2-m^2-\dfrac{r^2}{a^2}}{m^2\left(1-\dfrac{r^2}{m^2a^2}\right)}\right] & \text{for } r\le a \\ -\sigma_0 & \text{for } a\le r\le c \end{cases}$$

The total adhesive force is then given by,

$$F^D=-2\sigma_0 m^2a^2\left[\cos^{-1}(\frac{1}{m})+\frac{1}{m^2}\sqrt{m^2-1}\right]$$

The compression due to Dugdale adhesion is,

$$d^D=-\left(\frac{2\sigma_0 a}{E^*}\right)\sqrt{m^2-1}$$

and the gap at $r=c$ is,

$$h^D(c)=\left(\frac{4\sigma_0 a}{\pi E^*}\right)\left[\sqrt{m^2-1}\cos^{-1}\left(\frac{1}{m}\right)+1-m\right]$$

The net traction on the contact area is then given by $p(r)=p^H(r)+p^D(r)$ and the net contact force is $F=F^H+F^D$. When $h(c)=h^H(c)+h^D(c)=h_o$ the adhesive traction drops to zero.

Non-dimensionalized values of a,c,F,d are introduced at this stage that are defied as,

$$\bar{a}=\alpha a\ ;\ \bar{c}:=\alpha c\ ;\ \bar{d}:=\alpha^2 Rd;\ \alpha:=\left(\frac{4E^*}{3\pi\Delta\gamma R^2}\right)^{\frac{1}{3}};\ \bar{A}:=\pi c^2;\ \bar{F}=\frac{F}{\pi\Delta\gamma R}$$

In addition, Maugis proposed a parameter λ which is equivalent to the Tabor parameter μ. This parameter is defined as,

$$\lambda:=\sigma_0\left(\frac{9R}{2\pi\Delta\gamma E^{*2}}\right)^{\frac{1}{3}}\approx 1.16\mu$$

where the step cohesive stress σ_o equals to the theoretical stress of the Lennard-Jones potential,

$$\sigma_{th}=\frac{16\Delta\gamma}{9\sqrt{3}z_o}$$

Zheng and Yu suggested another value for the step cohesive stress,

$$\sigma_0 = \exp\left(-\frac{223}{420}\right) \cdot \frac{\Delta\gamma}{z_0} \approx 0.588 \frac{\Delta\gamma}{z_0}$$

to match the Lennard-Jones potential, which leads to,

$$\lambda \approx 0.663\mu$$

Then the net contact force may be expressed as,

$$\bar{F} = \bar{a}^3 - \lambda \bar{a}^2 \left[\sqrt{m^2 - 1} + m^2 \sec^{-1} m \right]$$

and the elastic compression as,

$$\bar{d} = \bar{a}^2 - \frac{4}{3} \lambda \bar{a} \sqrt{m^2 - 1}$$

The equation for the cohesive gap between the two bodies takes the form,

$$\frac{\lambda \bar{a}^2}{2} \left[(m^2 - 2) \sec^{-1} m + \sqrt{m^2 - 1} \right] + \frac{4 \lambda \bar{a}}{3} \left[\sqrt{m^2 - 1} \sec^{-1} m - m + 1 \right] = 1$$

This equation can be solved to obtain values of c for various values of a and λ. For large values of λ, $m \to 1$ and the JKR model is obtained. For small values of λ the DMT model is retrieved.

Carpick-Ogletree-Salmeron (COS) Model

The Maugis-Dugdale model can only be solved iteratively if the value of λ is not known a-priori. The Carpick-Ogletree-Salmeron approximate solution simplifies the process by using the following relation to determine the contact radius a:

$$a = a_0(\beta) \left(\frac{\beta + \sqrt{1 - F / F_c(\beta)}}{1 + \beta} \right)^{\frac{2}{3}}$$

where a_0 is the contact area at zero load, and β is a transition parameter that is related to by,

$$\lambda \approx -0.924 \ln(1 - 1.02\beta)$$

The case $\beta = 1$ corresponds exactly to JKR theory while $\beta = 0$ corresponds to DMT theory. For intermediate cases $0 < \beta < 1$ the COS model corresponds closely to the Maugis-Dugdale solution for $0.1 < \lambda < 5$.

Influence of Contact Shape

Even in the presence of perfectly smooth surfaces, geometry can come into play in form of the macroscopic shape of the contacting region. When a rigid punch with flat but oddy shaped face is carefully pulled off his soft counterpart, its detachment occurs not instantaneously but detachment fronts start at pointed corners and travel inwards, until the final configuration is reached which for macroscopically isotropic shapes is almost circular. The main parameter determining the adhesive strength of flat contacts occurs to be the maximum linear size of the contact.

Gear Failure and its Causes

Machinery breakdowns are often the result of gear failure. The cause of gear failure can stem from many issues, but the most common is due to wear. Any component within the machine is going to wear down from constant use. If it occurs with gears, it can have a detrimental impact on the equipment's operation.

Gear failures due to wear may occur for different reasons. The reasons listed below are some of the more common causes for gear failure:

- Moderate wear: This type of where leaves contacts patterns that show the metal has been affected in the addendum and the dedendum area. Issues with inadequate lubrication commonly cause it, but it may also be due to contamination in the lubrication as well.
- Excessive wear: This is wear that has continued to be a problem until a significant amount of material has been affected on the surfaces. You may see pitting on the surface with excessive wear, typically caused by not seeing the first wear early enough, and it continues to progress.
- Abrasive wear: In gears that are showing signs of abrasive wear, it may appear as radial scratch marks, grooves or some other identifier that would show contact is a problem. One of the most common reasons why abrasive wear takes place is due to foreign bodies in the lubrication. This issue could include metallic debris from the bearings or gear system, rust, sand or weld spatter. It is common in new systems before the filter can clean the system.
- Corrosive Wear: Chemical action deteriorates the surface of the gear, such as through acid, additives or moisture in the lubrication oil. As the oil breaks down, the chemicals that exist in the lubricant attack the surfaces. It tends to result in uniform, fine pitting on the two surfaces.
- Frosting: This issue usually shows up in the dedendum area of the driving gear.

The wear pattern gives a frosted appearance, which are many micro pits on the surface. Frosting is a common issue when the heat breaks down the lubrication film.

- Spalling: Although it is similar to severe pitting, the pits tend to be shallow and larger in diameter. Additionally, the area that is showing spalling does not tend to be uniform. It is a common problem when high contact stress exists.
- Pitting: A problem with pitting can be labeled as either initial, in which the surface is experiencing small pits to destructive, in which the pits are larger in diameter. Initial pitting may be a problem with the gears not fitting together properly. Destructive pitting is typically an issue with surface overload.
- Breakage: It is possible for the entire tooth or a piece of the tooth to break away. It often leaves evidence of the focal point of the fatigue that led to the break, which results from any number of issues, including high stress or excessive tooth loads.

Monitoring System of Gears

In order to minimise gearbox downtime and to avoid performance degradation, a practical and robust monitoring system is needed to provide early warnings of malfunction or possible damage. Extensive research efforts have been made to predict the health of gears using vibration and oil monitoring techniques. The present research focuses on the use of online monitoring techniques, which are useful for difficult-to-access gearboxes, to reduce unplanned gear failure and redundant planned maintenance.

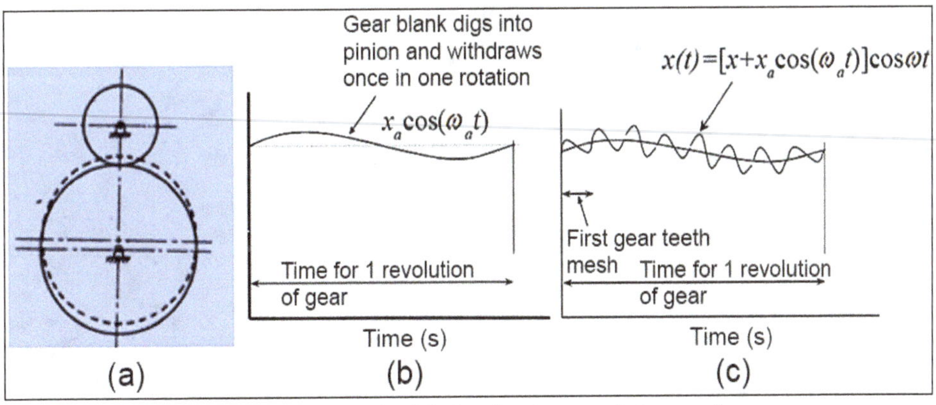

Eccentric gear pair: (a) sketch of eccentric gear pair; (b) signal due to eccentricity; (c) amplitude modulated signal.

Vibration analysis is the most commonly used monitoring technique to predict the condition of a gearbox. A gearbox contains a shaft, gears and bearings; therefore, the signals obtained from the vibration sensor contain: (i) primary rotation frequencies of the gear

(f_r) and pinion shafts and their harmonics; (ii) gear mesh frequency ($f_m = z.f_r$, where z is the number of teeth on the gear) and its harmonics; (iii) frequencies associated with the bearing supporting the pinion and gear shafts; (iv) sidebands of the gear mesh and gear mesh harmonics; (v) frequencies related to hobbing or cutting marks on the gear surface; and some random noise. The aim of the present research work is to find out the faults related to gears; therefore, frequencies directly related to gears shall be examined by extracting them with the help of signal pre-processing techniques. Lebold et al described various pre-processing techniques used in various vibration monitoring techniques.

Under non-ideal conditions, such as pitch errors, profile errors, variations in tooth spacing, misalignment, eccentricity and load variation, gear vibration signals contain amplitude and/or frequency modulation(s). For example, eccentricity causes amplitude modulation of a gear vibration signal due to the periodic variation in the depth of the mesh, which in turn varies the magnitude of the contacting force between a pair of teeth. It also results in frequency modulation of the time signal (but to a lesser degree), due mainly to the variation in the effective gear radius and the consequent variation in the angular speed of the mating gear. These gear defects can be easily detected by vibration monitoring, as shown in figures, provided all other frequency components have been filtered out. If these non-ideal conditions are detected at an earlier stage using vibration monitoring, then assembly corrections can be made and the expected life of gears can be improved significantly.

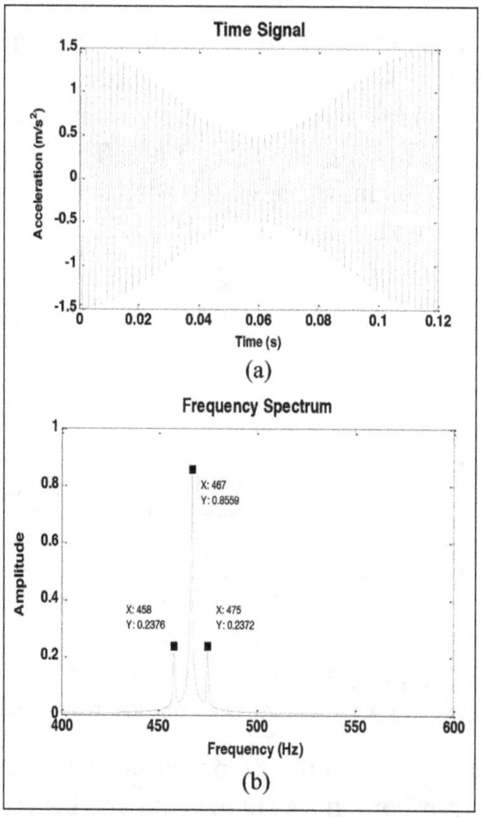

Signal model for amplitude modulation: (a) time plot; (b) frequency spectrum.

Faults, such as tooth tip breakage, fatigue cracking, pitting, etc., generated during gear operation weaken the gear teeth, reduce local mesh stiffness (when that faulty tooth is in mesh) and change the vibration behaviour by introducing an impulse during meshing of the affected tooth. The amount and duration of amplitude variation depends mainly upon the severity of the tooth defect and the contact ratio of the gear pair. If the tooth fault severity is small and the contact ratio is high, the resulting amplitude variation may not be seen distinctively on the vibration signal. The identification of such faults (bending crack, pitting, etc. produced during gear operation) at an early stage using vibration signals (i.e. sidebands) may be particularly difficult because these faults produce a very short duration of the modulation, which yields sidebands to extend over a broad frequency range but having very low amplitude. In such cases, oil analysis (lubricant and wear debris analyses) provides more reliable results in predicting the health of oil-lubricated gear pairs. Early detection of changes in lubricant condition and consistent monitoring of wear metal debris along with the lubricant provide greater insight into the actual condition of the gearbox. Oil analysis may be online or offline, as shown in figure.

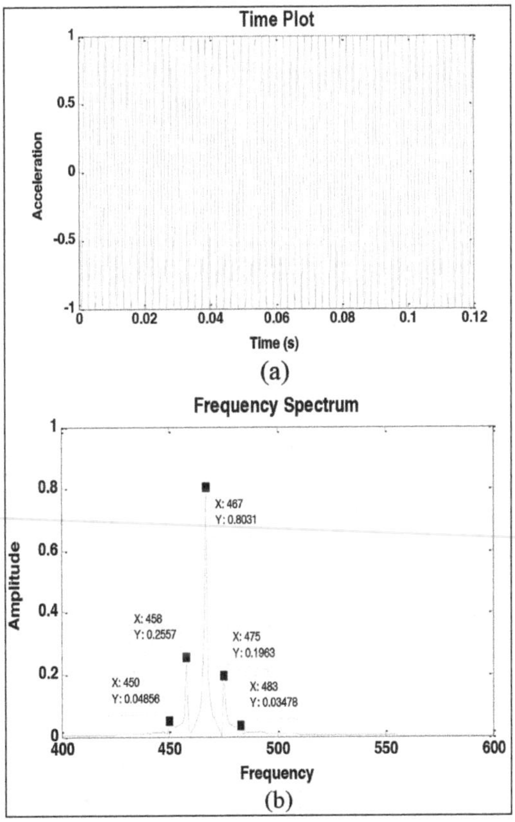

Signal for frequency modulation: (a) time plot; (b) frequency spectrum.

Although vibration monitoring is a well-accepted and widely used technique for detecting gear faults, most of the vibrations signals are noisy and different signal processing techniques are required to extract the required information. In the case of a complex system having so many components (i.e. a gearbox having bearings, seals, coupling, connection with the motor and connection with loading devices), the features extracted from the signal may give unreliable results. Also, operational effects can adversely impact the performance of vibration parameters and result in false alarms. On the other hand, oil analysis does not provide good results in the case of faults such as tooth cracks, eccentricity, misalignment, etc. Oil analysis is a better predictor of failures related to surface-like wear, pitting, scuffing, etc. But online oil analysis is ineffective for detecting large debris, as large debris having sufficient mass will settle down in the sump and cannot be captured by online oil analysis sensors. However, large debris (a particle size greater than 100 micrometers) can be monitored using a magnetic chip detector. Magnetic chip detectors provide a warning signal on the instrument panel indicating the presence of large-sized metallic chips. However, there is the availability of online metallic wear debris sensors, which can measure particles larger than 300 microns.

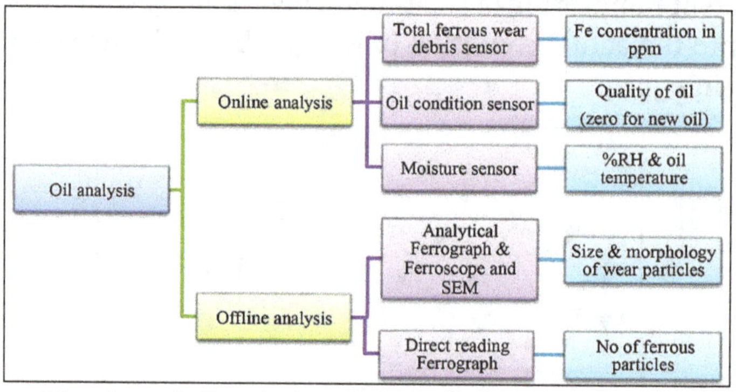

Flowchart of oil analysis techniques included in this study.

Generally, vibration and wear debris analyses are conducted independently. When used independently, they can only diagnose about 40-50% of faults. For example, vibration analysis indicates a gear fault (such as an increase in the vibration level at the gear mesh frequency), but does not provide the reason (i.e. a decrease in oil viscosity due to a mistake in topping-up the gear oil) for such an increase in gear vibration. Similarly, an increase in wear debris indicates a gear fault, but does not provide the reason (gear misalignment due to failure at the coupling level) for such an increase in the number of wear debris. In other words, oil and vibration analyses should be used jointly to monitor the root cause of gear failure. Without such a blend of technologies, the root of the gear problem may go undetected. Unfortunately, the two techniques are rarely combined to form an effective union. Two major reasons for such an ineffective union are:

- Vibration analysis has been dealt with by mechanical engineers, while oil analysis (i.e. moisture content, viscosity change, change in TAN number, etc.) resides with the lubrication team.
- Vibration analysis has been used as an online condition monitoring technique, while an oil analysis programme usually consists of submitting occasional samples to the laboratory, which is time consuming.

In the present research, vibration and online oil analyses have been treated as complementary to each other. Offline oil analysis was performed to ensure the correctness of online oil analysis.

Test-rig

Gearbox test set-up.

A gear test-rig, as shown in figure, has been developed for condition monitoring of a spur gear. This set-up consists of a single-stage gearbox driven by a shunt electric motor (30 kW DC) and a controller to regulate the motor speed in the range of 0-3000 r/min. The test gearbox consists of a pair of standard involute profiled spur gears and bearings (URB32306 bearing for the driving shaft and URB30307 bearing for the driven shaft). Lovejoy coupling connects the motor to the input shaft of the gearbox. The output shaft of the gearbox connects the shaft of the torque sensor through Lovejoy coupling. Torque (1-75 Nm) on the gears can be applied by an eddy current dynamometer (consisting of a LSG 2010 controller) coupled with the output shaft. To take care of the angular and linear misalignments, universal coupling was used between the dynamometer and the gearbox shaft.

Table: Specifications of the gearbox.

SR no.	Parameters	Pinion	Gear
1	Module	2	2
2	No of teeth	27	53
3	Pitch diameter	54	106
4	Outer diameter	58	110
5	Base diameter	50.7434	99.6074
6	Face width	33	33
7	Pressure angle	20°	20°
8	Contact ratio	1.697	1.697
9	Circular tooth thickness	3.1415	3.1415
10	Material	EN19	EN19

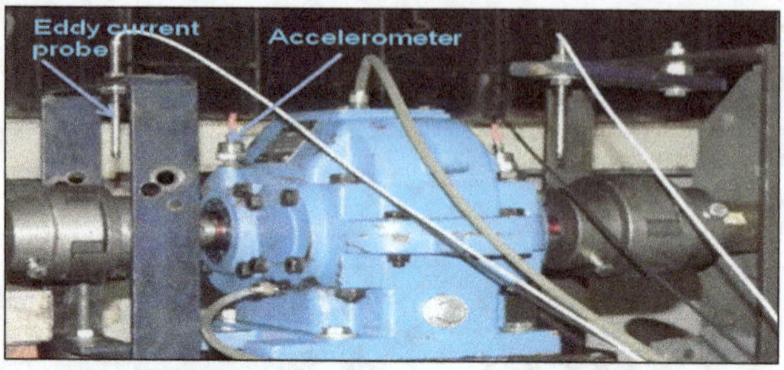

Gearbox with accelerometers and eddy current probes.

An accelerometer, mounted on the gearbox casing, has been used to measure the vibrations generated by the gear shown in figure. Eddy current probes, one for the input shaft and one for the output shaft of the gearbox, as shown in figure, provide a rotation count for these shafts. These data have been used in the time synchronous averaging process.

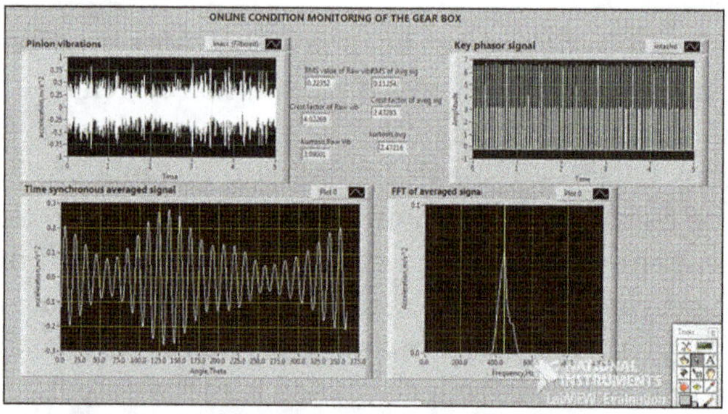

Front panel of online condition monitoring program.

Time synchronous averaging and FFT programs have been integrated with LabVIEW by using the Math script RT module. In addition, a program for calculating statistical parameters has been integrated into LabVIEW to estimate the RMS, crest factor and kurtosis values of the raw and averaged signals. Figure shows the front panel of the online condition monitoring LabVIEW program. The raw vibration signal of the gearbox is shown in figure. The keyphasor signal is shown in figure. The averaged signal is shown in figure and the FFT of the averaged signal is shown in figure. To gain confidence on the developed set-up, an initial experimental study on the gearbox was performed keeping a constant load of 15 Nm and a pinion speed of 600 r/min. For the tests, gear oil (API GL-4, SAE-80 W-90) has been used. The vibration signal from the healthy gearbox has been collected at the pinion (driver) and gear of the gearbox at a sampling frequency of 11 kHz. Figure shows the raw vibration signal from the pinion. This signal does not provide any information on the condition of the gearbox and time synchronous averaging is required to reduce the noise. Time synchronous averaged signals are shown in figure. This figure shows peaks, which are equal to the number of teeth on the pinion. This averaged signal does not have any transient signal, so the pinion is in a healthy condition.

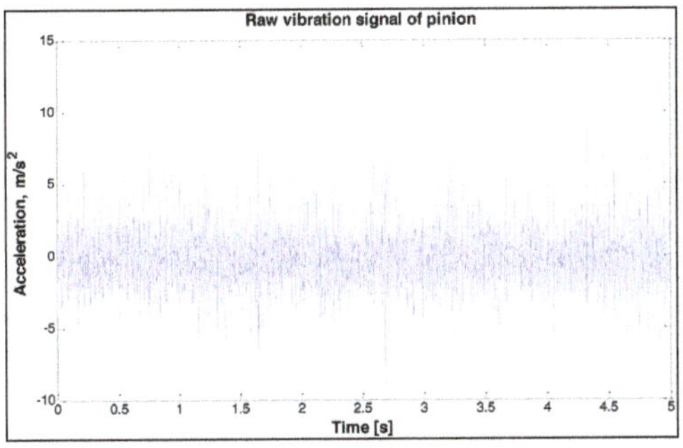

Raw vibration signal of the gearbox.

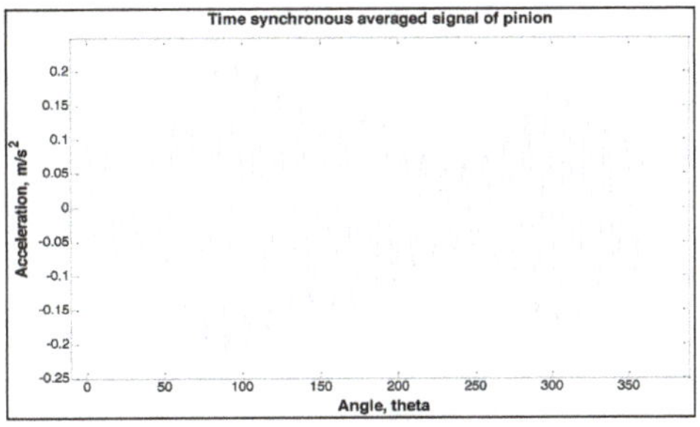

Time synchronous averaged signal of the pinion.

Figure shows the raw vibration signal of the driven gear. Figure is the time synchronous averaged signal of the gear. It contains 53 peaks, which are equal to the number of teeth on the gear. As it does not have any transient signals, the gear is in a healthy condition.

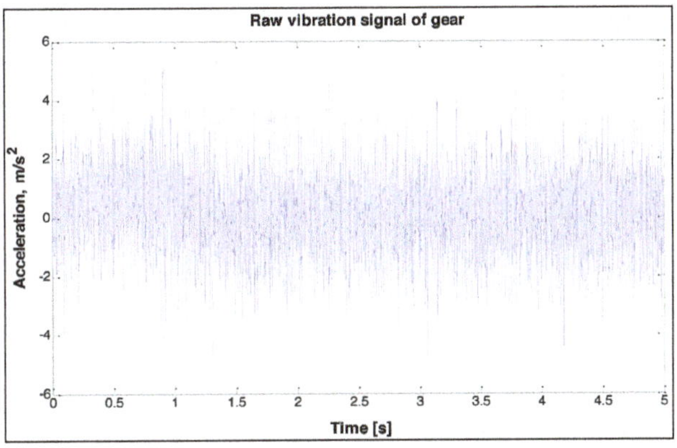

Raw vibration signal of the gear signal of the pinion.

The time synchronous averaged (TSA) signal of the pinion vibrations, filtered around the second harmonic of the gear mesh frequency, is shown in figure. This TSA signal amplitude is more than the TSA signal amplitude of the fundamental gear mesh frequency. This indicates the presence of misalignment in the gearbox.

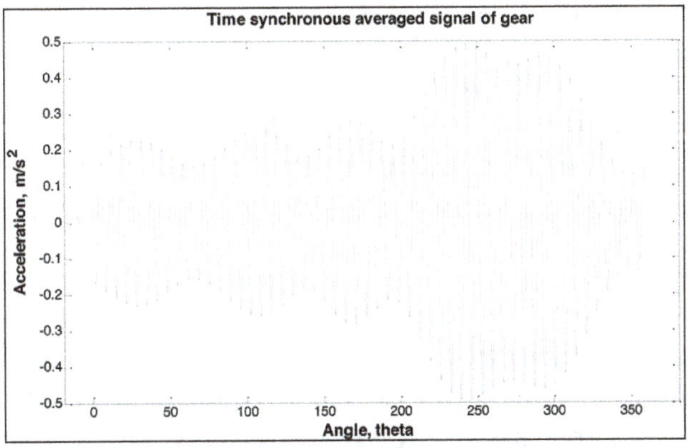

Time synchronous averaged signal of the gear.

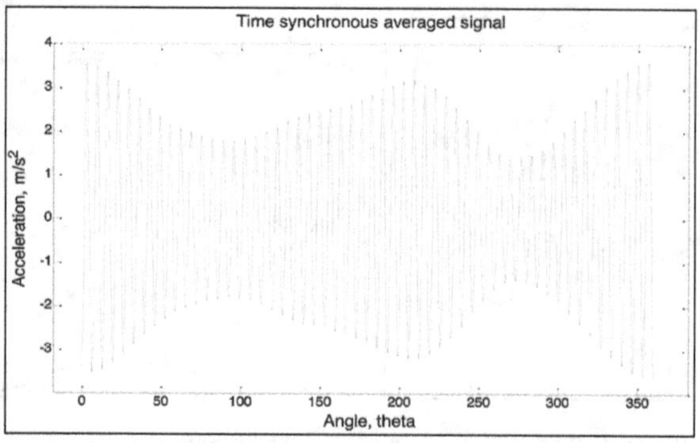

Time synchronous averaged signal of the pinion vibration signal filtered around the second gear mesh harmonic.

Similarly, the time synchronous averaged signal of the gear vibrations, filtered around the second harmonic of the gear mesh frequency, indicates the presence of misalignment in the gearbox. The gearbox vibration analysis reveals that gears present in the gearbox do not have any fault, but the gearbox has some misalignment. To minimise the misalignment, a common baseplate, as shown in figure, has been used.

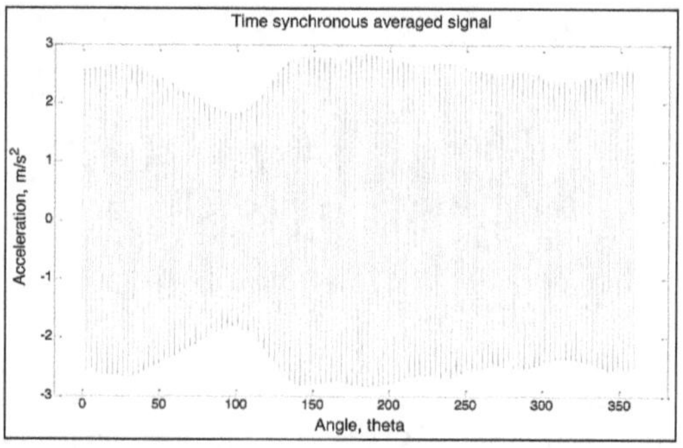

Time synchronous averaged signal of the gear vibrations filtered around the second gear mesh frequency.

To perform online oil analysis, a sensor suite, as shown in Figure, has been used. The sampling point, below the gear mesh where a high circulating flow predominates, has been selected. The oil from the gearbox passes through the sensors and returns back to the gearbox. The ANALEXrs online sensor suite consists of the following sensors:

- Total ferrous wear debris sensor: This sensor measures ferrous density, resulting from the wear debris within the lubricant, using a combination of magnetometry combined with smart algorithms, to provide data in parts per million (ppm) in the range of 0-2000 ppm. An increase in the ppm value of this sensor intimates the deterioration of the gears.

- Oil condition sensor: This sensor uses dielectric sensing combined with smart algorithms to provide trends. It checks the combined effect of TAN, a change in viscosity, and water ingress, and expresses quality as oil degradation on a 0-100 scale; generally zero is set as a reference for new oil. As the oil degrades, the oil quality number increases from the zero level.

- Moisture sensor: The moisture sensor measures relative humidity (0-100%) and oil temperature (–20 to 120 °C). Water in oil can increase the oxidation rate of the lubricant by more than ten times. The moisture sensor head allows only water molecules to penetrate its special polymer coating. The sensor monitors the dielectric property of the polymer layer, which has been affected by the water absorbed into the polymer. With increasing moisture, more and more water molecules are deposited on the polymer layer, thus increasing the dielectric constant of the material. This water content is reported as a percentage, indicating the humidity of the oil. In other words, this sensor uses a combination of thin-film capacitance sensors, combined with smart algorithms, to provide a temperature and % RH value.

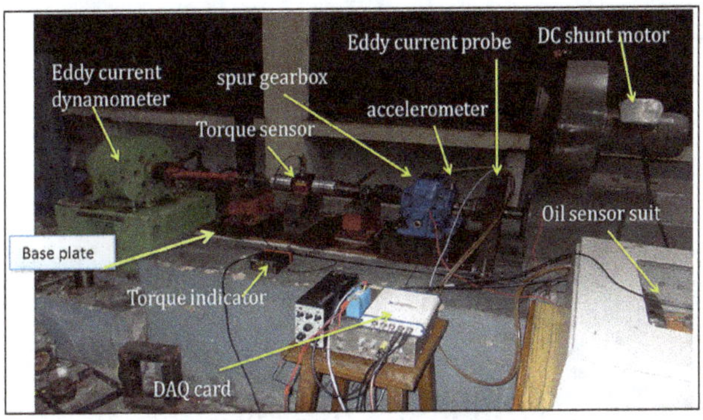

Modified gearbox test set-up.

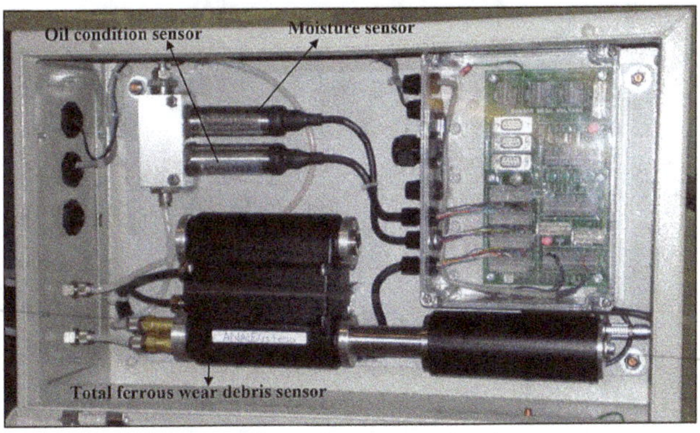

Online oil sensor suite.

Results and Discussion

Experiments were carried out on the developed test-rig. The input shaft was rotated at different rotational speeds (300 r/min, 394 r/min and 400 r/min) and various torque loads (7 Nm, 15 Nm and 20 Nm) were applied on the gear output shaft. Vibration- and oil-related data were collected for each operating condition. To accelerate the gear failure rate, hydrochloric acid was added and readings for three cases (Case: no acid; Case: two drops of acid; and Case: four drops of acid) were noted down. Acid increases the TAN number of the oil, which deteriorates the oil quality and accelerates the rate of gear failure.

Vibration Results

Data from the accelerometer, mounted on the bearing housing of the gearbox case, were recorded for a one second duration using a DAQ card and computer.

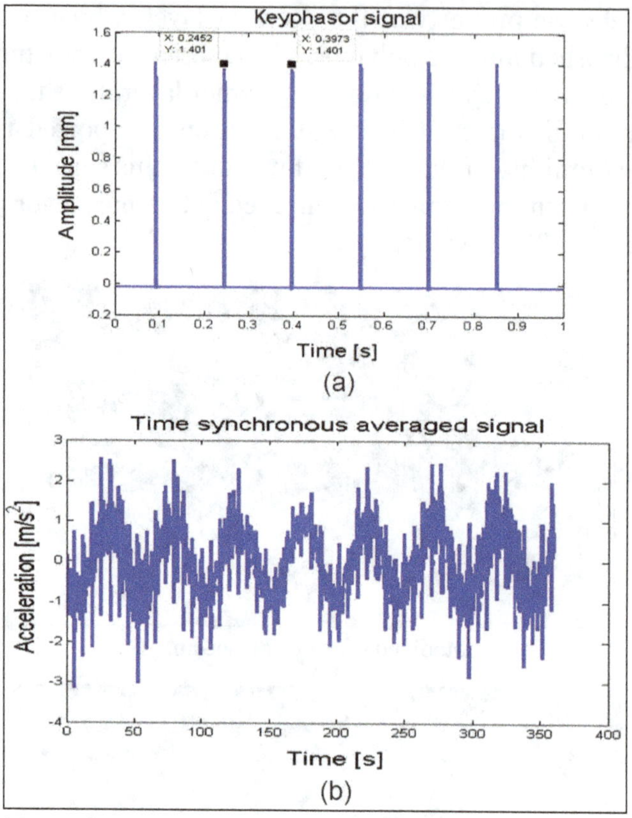

(a) Keyphasor signal for speed = 394 r/min, torque = 20 Nm;
(b) TSA of signal for speed = 394 r/min, torque = 20 Nm.

Figure above shows the key phasor signal, which gives the rotation of the gear shaft. The rotational frequency from the signal is calculated as 6.574 Hz. Figure shows the total number of peaks (53) in one revolution, which should be the total number of teeth on the gear. Figure above shows the FFT of the TSA signal, as shown in figure.

Figure clearly shows a lesser magnitude of the second harmonic (694.3) of the GMF compared to the magnitude of the GMF (348.5) itself. This means a misalignment-related problem has been corrected. But the magnitude at the GMF is relatively lesser compared to the magnitude of other frequencies (ie 447.1, 545.7 and 742.9 Hz). This means vibration monitoring analyses of the whole of the system and compared to gears in other components of the test set-up (i.e. the bearing of the techgenerator) experience relatively larger acceleration. To determine the faults related to gears, all the faults and corresponding characteristics must be known in advance.

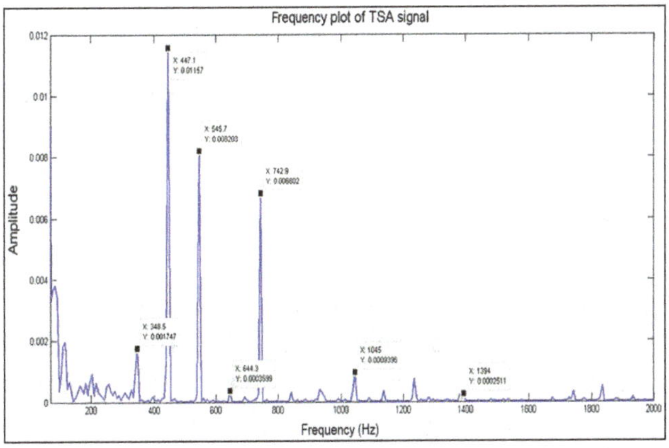

FFT of TSA signal for speed = 394 r/min, torque = 20 Nm.

For example, it is clear that the gear provides a higher value of acceleration at its harmonics and use of a band-pass filter around the gear mesh frequency and its harmonic shall provide better results, as shown in figure. But knowing all frequencies related to faults often puts limits on online health monitoring. In addition, sometimes spurious signals at the rotational frequency and its harmonics are obtained. After observing such signals, subsequently opening the gearbox reveals just a larger compressed debris on the gear tooth surface. In such cases, oil monitoring provides the required information of gear health.

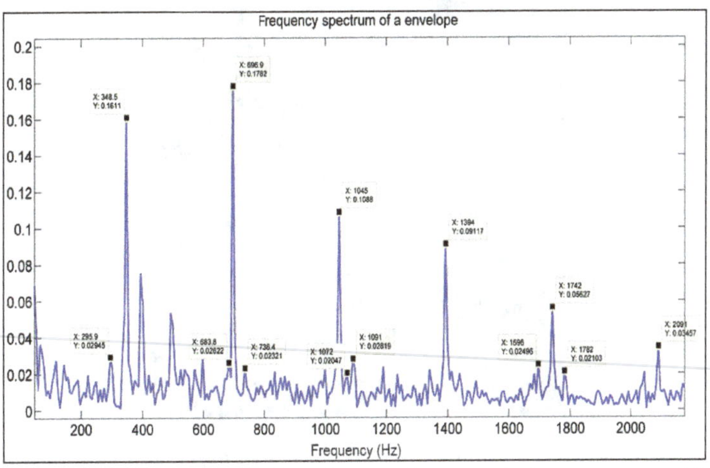

Enlarged view of FFT of envelope for speed = 394 r/min, torque = 20 Nm.

Online Oil Analysis

Data were taken from the online sensor suite (for operating speed = 300 r/min, torque = 7 Nm) to monitor the trend of Fe concentration (in ppm) and oil quality number. To see the faster failure rate, an accelerated test was performed by adding hydrochloric acid to the oil to increase the wear rate and deteriorate the oil quality. Wear rates were compared for different speeds (300 r/min and 400 r/min) and torque conditions (no load, 7 Nm and 15 Nm).

Table: Comparison of Fe concentration (in ppm) for speed = 300 r/min, torque = 7 Nm.

SR No.	Case	Maximum ppm reading	Mean ppm reading
Case No 1	With no acid	130	117
Case No 2	With two drops of acid	140	125
Case No 3	With four drops of acid	168	150.5

Figure shows Fe concentration in ppm for case 1, case 2 and case 3. As listed in table, the wear rate increases by adding the acid to the lubricating oil. At the same speed, on increasing the load from 7 Nm to 15 Nm, an increase in maximum wear debris from 168 ppm to 210 ppm is observed, as shown in figure. Figure shows the result for Fe concentration in ppm r/min for the no load condition, but with an increased speed to 400 r/min. In this Fe (ppm), concentration is reduced from 168 to 160. These trends clearly indicate an increase in wear rate with an increase in load and speed.

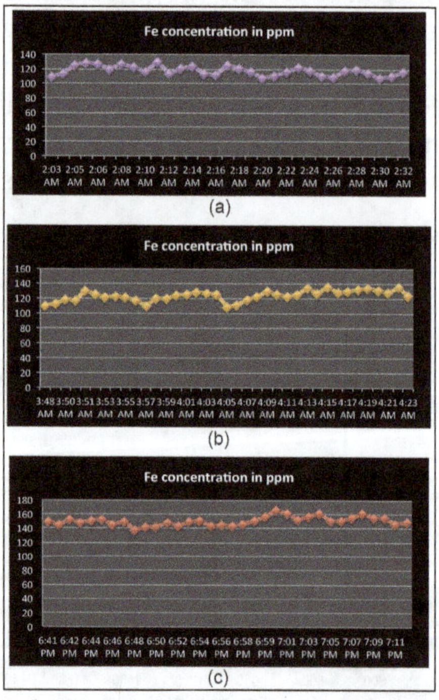

Fe concentration (ppm) for speed = 300 r/min, torque = 7 Nm: (a) Case 1: Oil without adding acid; (b) Case 2: Two drops of acid is added to the oil; (c) Case 3: Four drops of acid in the lubricating oil.

To rely on these online analysis results, offline analysis was carried out. Offline oil analysis was performed by three techniques: (1) direct reading ferrograph; (2) analytical ferrograph; (3) scanning electron microscope.

To acquire the results of the direct reading ferrograph, the first instrument was calibrated with fixer oil (tetrachloroethylene). Sample 1 was prepared by mixing 1 ml fixer + 1 ml oil (extracted from the gearbox). The results of DL (particles greater than 5 micrometer) and DS (particles less than 5 micrometers) were checked. As the values of DL and DS were larger than 90, the instrument response becomes non-linear due to the particles piling on the top of one another so that less light is attenuated. To rectify this, the sample was diluted and the procedure was repeated. The results are listed in table.

Table: Results of direct reading ferrograph.

SR No	DL reading	DS reading
Sample 1	127.6	100.5
Sample 2	80.6	54.8
Sample 1: 1 ml oil + 1 ml fixer (1:1 dilution) (total 2 ml of sample). Sample 2: 50% of the base sample (already diluted oil) + 1 ml of fixer (total 2 ml).		

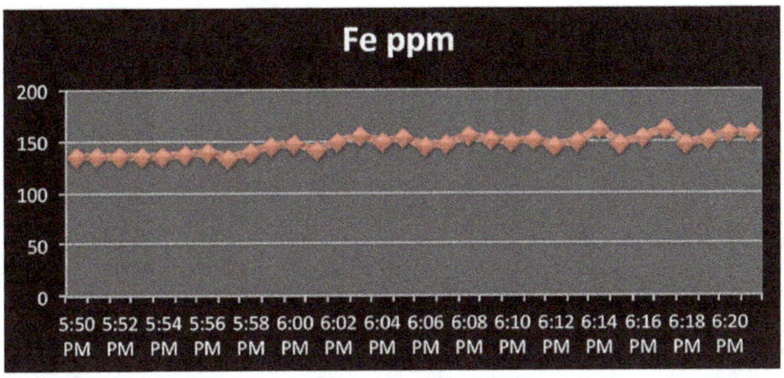

Fe concentration (ppm) for four drops of acid in oil, speed = 400 r/min, no load.

The DL and DS readings have to be multiplied by the dilution factor to get the correct readings. In this case, the dilution factor was 2 (because of 50% dilution). So, the correct reading of the oil sample is:

$$DL = 161.2$$

$$DS = 109.6$$

DL and DS values are used to calculate the wear severity index, ID, where:

$$ID = (DL + DS)(DL - DS) = 13973.28$$

- In normal rubbing wear, the majority of the particles are small, so DS will be comparable with DL, and ID will be small.
- If the wear regime is more severe, DL will be large compared to DS, i.e. ID increases with increasing severity of wear.

These results confirm the significant wear of the gear pair due to the addition of acid in the oil.

References

- Spur-gears-what-are-they-and-where-are-they-used: motioncontroltips.com, Retrieved 19 April, 2020
- Eight-common-causes-of-gear-failure: plantengineering.com, Retrieved 26 January, 2020
- Popov, Valentin L., 2010, Contact Mechanics and Friction. Physical Principles and Applications, Springer-Verlag, 362 p., ISBN 978-3-642-10802-0
- L Parvathareddy, 'Online condition monitoring of spur gears', M.Tech thesis, Mechanical Engineering Department, IIT Delhi, 2010

PERMISSIONS

All chapters in this book are published with permission under the Creative Commons Attribution Share Alike License or equivalent. Every chapter published in this book has been scrutinized by our experts. Their significance has been extensively debated. The topics covered herein carry significant information for a comprehensive understanding. They may even be implemented as practical applications or may be referred to as a beginning point for further studies.

We would like to thank the editorial team for lending their expertise to make the book truly unique. They have played a crucial role in the development of this book. Without their invaluable contributions this book wouldn't have been possible. They have made vital efforts to compile up to date information on the varied aspects of this subject to make this book a valuable addition to the collection of many professionals and students.

This book was conceptualized with the vision of imparting up-to-date and integrated information in this field. To ensure the same, a matchless editorial board was set up. Every individual on the board went through rigorous rounds of assessment to prove their worth. After which they invested a large part of their time researching and compiling the most relevant data for our readers.

The editorial board has been involved in producing this book since its inception. They have spent rigorous hours researching and exploring the diverse topics which have resulted in the successful publishing of this book. They have passed on their knowledge of decades through this book. To expedite this challenging task, the publisher supported the team at every step. A small team of assistant editors was also appointed to further simplify the editing procedure and attain best results for the readers.

Apart from the editorial board, the designing team has also invested a significant amount of their time in understanding the subject and creating the most relevant covers. They scrutinized every image to scout for the most suitable representation of the subject and create an appropriate cover for the book.

The publishing team has been an ardent support to the editorial, designing and production team. Their endless efforts to recruit the best for this project, has resulted in the accomplishment of this book. They are a veteran in the field of academics and their pool of knowledge is as vast as their experience in printing. Their expertise and guidance has proved useful at every step. Their uncompromising quality standards have made this book an exceptional effort. Their encouragement from time to time has been an inspiration for everyone.

The publisher and the editorial board hope that this book will prove to be a valuable piece of knowledge for students, practitioners and scholars across the globe.

INDEX

www.ingramcontent.com/pod-product-compliance
Lightning Source LLC
LaVergne TN
LVHW080621210926
848485LV00003B/192